ANCIENT AND MODERN MATHEMATICS

1)- ANCIENT PROBLEMS
2)- PARTIAL PERMUTATIONS

DAT PHUNG TO

Order this book online at www.trafford.com
or email orders@trafford.com

Most Trafford titles are also available at major online book retailers.

© Copyright 2012 Dat Phung To.
All rights reserved. No part of this publication may be reproduced, stored in a retrieval system, or transmitted, in any form or by any means, electronic, mechanical, photocopying, recording, or otherwise, without the written prior permission of the author.

Printed in the United States of America.

ISBN: 978-1-4669-0094-3 (sc)
ISBN: 978-1-4669-0093-6 (hc)
ISBN: 978-1-4669-0095-0 (e)

Library of Congress Control Number: 2012909213

Trafford rev. 02/07/2013

 www.trafford.com

North America & international
toll-free: 1 888 232 4444 (USA & Canada)
phone: 250 383 6864 ♦ fax: 812 355 4082

ForeWord Reviews
Clarion Review

EDUCATION

Ancient and Modern Mathematics: The Partial Permutations
Dat Phung To
Trafford
978-1-4669-0094-3
Five Stars (out of Five)

In *Ancient and Modern Mathematics*, Dat Phung To offers a refreshing postulation that mathematics can be appreciated on a more fundamental level than how it is often presented in these modern times of advanced-function calculators and whizbang computers. Dat Phung To is, in his own words, "the man who loves mathematics." And while he surely is not the only one, he does prove true to his self-description throughout the pages of this gem of a book.

Ancient and Modern Mathematics is divided into two sections, "Arithmetic and Geometric Problems" and "The Partial Permutations," two of no doubt many areas of mathematics of which the author has made a study for his own enjoyment.

The arithmetic problems include classic challenges such as "The Sum of Rice in Sixtyfour Squares of Chessboard," in which a man in the Middle Ages invents the game of chess and shows the king how to play it. To reward the man, the king grants him a wish. Having always lived in poverty, the man asks for an amount of rice to be derived from placing one grain on the first square on the chessboard, two on the second, and so forth, each time doubling the number of grains of rice over the sixty-four squares. The king's mathematician makes the calculation—without the advantage of logarithmic tables or an electronic calculator, of course. Not only is the final sum arrived at as astounding as the reader might expect, but Dat Phung To's explanation of the most efficient calculation the king's mathematician might have used is presented precisely and cleary.

The geometric problems presented are also classics. The first is whether a triangle having two angles bisected by line segments of equal length must be an isosceles triangle. The discussion then moves on to the euclidean theory that a straight line and a circle intersect at two points. The latter may seem intuitively obvious, and readers are told that Euclid considered it so. But the author is not satisfied until he can prove the case, and he does. His hand-drawn diagrams add authenticity to his work and again demonstrate the simple beauty of pure, diligent mathematical work.

The second section of the book begins with a look at partial permutations. The author clearly lays out the main rules and then goes on to develop his own corollary to a conventional permutations theorem, which he then applies to further expansions. More handwritten charts appear, meticulously written and very readable.

There are a few typos in the text and a couple of other flaws, such as an occasional missing word, none of which alters the book's readability. Another round of editing would eliminate these errors for a second edition. And, the reader should be warned that the word "nominator" is used in place of "numerator," a rare but not actually incorrect usage. The author might consider altering the subtitle to include a reference to the first section of the book as well as the second. Finally, an overall table of contents, rather than one just for each section, would be helpful.

This is without doubt the work of an inspired man. Any math teacher, especially of algebra (heavily relied on in the first several problems), geometry, or discrete mathematics should have this thought-provoking book in his or her personal library and can recommend its reading to students as preparation for some interesting class discussion. Additionally, instructors of computer science might challenge their students to convert Dat Phung To's work into ditigal algorithms, just for the fun of it.

Tricia Morrow

Ancient and Modern Mathematics: The Partial Permutations
Dat Phung To
Trafford Publishing, 223 pages,
(paperback) $15.86, 978-1-4669-0094-3
(Reviewed: January 2013)

In a market glutted with mathematics textbooks, *Ancient and Modern Mathematics* is a refreshing approach to mathematical problems that have been around since ancient days. Author Dat Phung To says, "Studying the ancient problems is my great pleasure; pursuing solving them is my favorite hobby."

He gives readers more than a cursory look at these iconic mathematical challenges. At the same time, he convinces readers that the solutions can be determined through the use of paper and pencil rather than programmable calculators. After all, if the ancient mathematicians could do it sans electronics, so can the modern practitioner.

In the book's first half, "Arithmetic and Geometric Problems," the author works 11 classic arithmetic problems before delving into geometry. The first two, "A Basket of Eggs" and "The Sum of Rice in Sixty-four Squares of Chessboard," will be familiar to most mathematicians, even students. One problem requires only a firm grasp of algebra. The other applies binomial expansion using Chinese mathematician Chu Shih-Chieh's triangle of coefficients, which evidently predates Pascal's by some 360 years.

In the book's second half, "The Partial Permutations," the author introduces the relevant rules, definitions, and symbols in a logical, reader-friendly manner. He then navigates through partial permutations in a traditional way and through his own unique method of expansion.

The book is charmingly laced with bits of history, but it's the math, the meat of the book, that makes it a worthy addition to any mathematician's library. By offering multiple solutions to each problem, the author aims to offer readers some alternative thinking. He hits the mark again and again. His diagrams befit his sensible explanations, and his charts are nothing less than a labor of love.

Students of algebra, geometry, and discrete math will appreciate *Ancient and Modern Mathematics*, and teachers should consider adding it as a supplement to their standard textbook fare. One hopes there are more books to come from this "man who loves mathematics."

Also available in hardcover.

Preface

Dear readers, I am Dat Phung To, the man who loves mathematics. Throughout my life, I used most of my spare time to research mathematics, mostly the ancient arithmetic problems. The more I worked on them, the more I admired the ancient mathematicians who invented them. Unlike humans today who live in favorable conditions with advanced technological products, ancient people had to live in the reverse situation, but they could still achieve their great works. Therefore, I consider them to be my admirable teachers. Studying the ancient problems is my great pleasure; pursuing solving them is my favorite hobby.

I am writing this text to do something that might be interesting for those who love mathematics. My textbook consists of two parts. In the first part, there are fourteen ancient problems. In the second part, I introduce the partial permutations theory. The problems in the first part are set up in order from the easiest to the toughest. My own method improves and solves them.

I unexpectedly discovered the partial permutations in the second part while I searched the base e in the website, in which I fortunately read the Bernoulli problem about the derangement of the hats that brought me the ideas I used to develop the partial permutations theory. This theory proves that $0! = 1$ is no longer a convention but a corollary of Theorem 1. It astonishes us. The theory also provides us a specific and unique method to write down the whole expansion of $nPn = n!$ into single permutations with n being a finite number.

I hope the subject matters in my textbook do not coincide with those of other authors worldwide. Also, I hope my work will provide some pleasure to those who love mathematics, or at least it will help them to not despair.

Dat Phung To
April 30, 2011

Part I

Ancient Problems

Contents

Chapter 1

Problem 1: A Basket of Eggs..................5

Problem 2: The Sum of Rice in Sixty-four Squares of Chessboard..................12

Chapter 2

Problem 3: The Hundred Fowls (1)..................22

Problem 4: The Hundred Fowls (2)..................25

Problem 5: The Thousand Fowls..................28

Chapter 3

Problem 6: The Problem of Sunzi..................38

Problem 7: The Improved Problem of SUNZI to Inspect Battalion..................43

Problem 8: The Improved Problem of SUNZI to Inspect Regiment..................55

Problem 9: The Improved Problem of Sunzi to Inspect Division..................63

Problem 10: The Improved Problem of Sunzi to Inspect Corps..................75

Problem 11: The Improved Problem of Sunzi to Inspect a Group of People..................87

Chapter 4

Problem 12: The Triangle with Two Equal Bisectors............100

Problem 13: A Straight Line and a Circle Intersect at
　　　　　Two Points..109

Problem 14: Two Circles Have Only Two
　　　　　Common Points...125

Chapter I

Problem I: A Basket of Eggs

This problem might be from either ancient Vietnam or China. It is both funny and tricky. A peasant brought a basket of eggs to the outdoor market. A moment later, the first customer came to buy some eggs, but the peasant told him: "I would like to sell half of the eggs in the basket and a half of an egg. If you agree, you will get a great discount." This customer accepted this condition, and he brought the eggs home. Then the second customer came, and he did the same as the first. After that, the third customer came, and again, he did the same manner as the first two customers. After that time, the basket ran out of eggs.

Question 1	Question 2
How many eggs are in the peasant's basket?	In general, if n customers took their turn to buy the eggs with the same manner above and then the basket ran out of eggs, how many eggs are in the basket?

Solution I: Question I

First, we set x to be the numbers of eggs in the basket. Then we follow the three steps as follows. The first customer purchases the following number of eggs:

$$S_1 = \frac{x}{2} + \frac{1}{2} = \frac{x+1}{2}$$

The remaining eggs are:

$$R_1 = x - \frac{x+1}{2} = \frac{x-1}{2}$$

The second customer purchases the following number of eggs:

$$S_2 = \frac{1}{2}\left(\frac{x-1}{2}\right) + \frac{1}{2} = \frac{x+1}{2^2}$$

The remaining eggs are:

$$R_2 = \frac{x-1}{2} - \frac{x+1}{2^2} = \frac{x-3}{2^2}$$

The third customer purchases the following number of eggs:

$$S_3 = \frac{1}{2}\left(\frac{x-3}{2^2}\right) + \frac{1}{2} = \frac{x+1}{2^3}$$

The remaining eggs are :

$$R_3 = \frac{x-3}{2^2} - \frac{x+1}{2^3} = \frac{x-7}{2^3}$$

According to the hypothesis, the basket runs out of eggs after the third purchase. Therefore, we have:

$$R_3 = \frac{x-7}{2^3} = 0$$

Hence: $x = 7$

Solution I: Question 2

Let x be the number of eggs in the basket. Then we observe the results of the question 1 as follows. The first purchase is:

$$S_1 = \frac{x+1}{2^1} \qquad R_1 = \frac{x-1}{2^1} = \frac{x-(2^1-1)}{2^1}$$

The second purchase is:

$$S_2 = \frac{x+1}{2^2} \qquad R_2 = \frac{x-3}{2^2} = \frac{x-(2^2-1)}{2^2}$$

The third purchase is:

$$S_3 = \frac{x+1}{2^3} \qquad R_3 = \frac{x-7}{2^3} = \frac{x-(2^3-1)}{2^3}$$

In these formulas above, the only one different factor is the exponent of 2. The first purchase is 2^1. The second purchase is 2^2. The third purchase is 2^3. So we can apply the recurrent reasoning method to find out the formulas of sold eggs and the remaining eggs at the nth purchase. First, we suppose that the formulas of the sold and remaining eggs are still true at the $(n-1)^{th}$ purchase. Then we must prove that these formulas are also true at the nth purchase. We suppose:

$$S_{(n-1)} = \frac{x+1}{2^{(n-1)}} \quad \text{and} \quad R_{(n-1)} = \frac{x - \left[2^{(n-1)} - 1\right]}{2^{(n-1)}}.$$

Thus, the nth customer comes to purchase eggs, we have:

$$S_n = \frac{1}{2}\left[\frac{x - \left(2^{n-1} - 1\right)}{2^{n-1}}\right] + \frac{1}{2}$$

$$= \frac{x - \left(2^{n-1} - 1\right)}{2^n} + \frac{2^{n-1}}{2^n}$$

$$S_n = \frac{x+1}{2^n} \qquad (1)$$

The remaining R_n is:

$$R_n = \frac{x - \left(2^{n-1} - 1\right)}{2^{n-1}} - \frac{x+1}{2^n}$$

$$= \frac{2\left[x - \left(2^{n-1} - 1\right)\right]}{2^n} - \frac{x+1}{2^n}$$

$$R_n = \frac{x - \left(2^n - 1\right)}{2^n} \qquad (2)$$

The formulas (1) and (2) are proved, so it will be true at any numbers of purchases. According to the hypothesis, the basket runs out of eggs after the nth purchase, so we have:

$$R_n = \frac{x-(2^n-1)}{2^n} = 0$$

Thus: $x = 2^n - 1$

Solution 2: Question I

The first customer purchases the following number of eggs:

$$S_1 = \frac{x+1}{2^1} \text{ (the result of the first solution).}$$

The second customer purchases the following number of eggs:

$$S_2 = \frac{x+1}{2^2}$$

The third customer purchases the following number of eggs:

$$S_3 = \frac{x+1}{2^3}$$

Because the basket runs out of eggs after the third purchase, if we add up all these three purchases, we have:

$$\frac{x+1}{2^1} + \frac{x+1}{2^2} + \frac{x+1}{2^3} = x \quad (1)$$

If we set 2^3 to be the common denominator, the equation (1) becomes:

$$\frac{2^2(x+1) + 2(x+1) + (x+1)}{2^3} = x$$

Or we can rearrange:

$$(4x+4)+(2x+2)+(x+1) = 2^3 x$$

or

$$7x + 7 = 8x$$

Therefore, we have:

$$x = 7$$

Solution 2: Question 2

For the general case, we add all the sold eggs from the first to the nth purchase:

$$\frac{x+1}{2^1} + \frac{x+1}{2^2} + \frac{x+1}{2^3} + \dots + \frac{x+1}{2^n} \quad \text{(results proved in solution 1)}$$

Because the basket runs out of eggs after the nth purchase, that means the sum above is equal to x eggs in the basket. Thus, we have:

$$\frac{x+1}{2^1} + \frac{x+1}{2^2} + \frac{x+1}{2^3} + \dots + \frac{x+1}{2^n} = x$$

We recognize that all the terms on the lefthand side of the equation above form a geometric series of which the first term is $\frac{x+1}{2}$ and the common ratio is $\frac{1}{2}$. So we have the sum of the series as:

$$S = a_1 \frac{1-r^n}{1-r} \quad (1)$$

If we substitute $s = x$, $a_1 = \frac{x+1}{2}$, and $r = \frac{1}{2}$, we have:

$$x = \frac{x+1}{2} \cdot \frac{1-\left(\frac{1}{2}\right)^n}{1-\frac{1}{2}}$$

$$= \frac{x+1}{2} \cdot \left(\frac{\frac{2^n - 1}{2^n}}{\frac{1}{2}}\right)$$

$$x = (x+1) \cdot \frac{(2^n - 1)}{2^n}$$

Or we can write:

$$2^n \cdot x = (x+1) \cdot (2^n - 1)$$
$$2^n \cdot x = 2^n \cdot x - x + 2^n - 1$$

Therefore, we deduce:

$$x = 2^n - 1.$$

Note: The original problem only mentioned the particular case of three customers, not general case of n customers.

Problem 2: The Sum of Rice in Sixty-four Squares of Chessboard

This legendary problem might be from either Persia, India, or China in the Middle Ages. Its story has been spread orally throughout the Asiatic from generation to generation for many centuries. An old man invented a chess game. Then he brought it to the king and showed him how to play it. The king was very interested and appreciated this great work.

To reward the man, the king asked him, "What do you wish for? I could make your dream come true."

The old man replied, "Your Majesty, I am very poor, and I have lived in misery, so I only need enough rice to eat for the rest of my life."

The king said, "Sure, I can do it, but how much rice do you want?"

The old man slowly replied, "Your Majesty, as you know, the chessboard consists of sixty-four squares. I would like you to put one grain of rice in the first square, then two grains in the second, and then four grains in the third. The process is to double the amount of rice in the next square until the last one. If you don't mind, give me the total amount of rice in the chessboard."

Ancient and Modern Mathematics

The king smiled at this modest wish and said to the old man, "Of course I do. Tomorrow you'll come to the royal granary to collect them."

Then the king immediately gave the order to the expert mathematicians in his cabinet to solve this problem for the granary keeper.

Question 1	Question 2
How many grains of rice are in the total of sixty-four squares?	If the weight of eight grains of rice is about one gram, how many tons did the old man get?

Note: Remember, at that time, humans had no logarithmic tables, electronic calculators, or modern mathematic formulas. So to solve this problem, that man could only use simple formulas that might have existed at that time, but his result still had an accuracy of five significant digits.

Solution: Question I

There are two methods used to calculate the sum of rice:
1. The old man might use this method:
    a. We add the sum of rice in the first two squares (squ_1 and squ_2):

$$Squ_1 + squ_2 = Squ_3 - 1$$

or

$$2^0 + 2^1 = 2^2 - 1 = 3$$

b. We add the sum of rice in the first three squares:

$$Squ_1 + squ_2 + squ_3 = squ_4 - 1$$

or

$$2^0 + 2^1 + 2^2 = 2^3 - 1 = 7$$

c. We do the same manner for the first four squares:

$$Squ_1 + squ_2 + squ_3 + squ_4 = squ_5 - 1$$

or

$$2^0 + 2^1 + 2^2 + 2^3 = 2^4 - 1 = 15$$

d. This process will reach the sum of rice in the sixty-four squares:

$$Squ_1 + squ_2 + squ_3 + squ_4 + \ldots + squ_{64} = squ_{65} - 1$$

(squ_{65} for reference only)

or

$$2^0 + 2^1 + 2^2 + 2^3 + \ldots + 2^{63} = 2^{64} - 1$$

Thus, the total sum of rice is:

$$s = 2^{64} - 1$$

2. Because the process is to double the amount of rice in the next square until the last (sixty-fourth) square, we can write the sequence as below:

ANCIENT AND MODERN MATHEMATICS

$$squ_1 \ squ_2 \ squ_3 \ squ_4 \ \ldots \ squ_{64}$$
$$2^0 \ 2^1 \ 2^2 \ 2^3 \ \ldots \ 2^{63}$$

We recognize that the total amount of rice is the sum of rice in sixty-four squares above. This amount is the sum of a geometric series consisting of sixty-four terms, the first term is $2^0 = 1$, and the common ratio is 2. Thus, we deduce the total amount of rice by the formula:

$$s = a_1 \cdot \frac{1-r^n}{1-r}$$
$$= 1 \cdot \frac{1-2^{64}}{1-2}$$
$$s = 2^{64} - 1$$

Solution: Question 2

To convert $s = 2^{64} - 1$ grains of rice into weight, we first ignore the amount 1. We only consider the great amount (2^{64}) as the sum of rice. Then we develop $s = 2^{64}$ into decimal numeration system as below: $s = 2 \times 2 \times 2 \times 2 \times 2 \times \ldots \times 2$ sixty-four times. If we calculate 2^{64} in the manner above without the help of logarithmic tables or electronic calculators, it will take a lot of extremely hard work and time and result in easy errors. Therefore, we try to find out such a way that allows the calculating to be easier and to take less time. First, we look at the consecutive terms 2^n of the sequence below:

2^0	2^1	2^2	2^3	2^4	2^5	2^6	2^7	2^8	2^9	2^{10}
"	"	"	"	"	"	"	"	"	"	"
1	2	4	8	16	32	64	128	256	512	1,024

We recognize that $2^{10} = 1,024$ can form a binomial:

$$2^{10} = 10^3 + 24$$

Thus, the total grains of rice can be written:

$$s = 2^{64} = 2^4 \times 2^{10 \times 6} = 16(10^3 + 24)^6$$

If we set $A = 10^3$ and $B = 24$, the binomial becomes:

$$s = 16(A + B)^6 = 16(A + B)^{(2 + 2 + 2)} = 16(A + B)^{2 \times 3}$$

Because the weight of eight grains of rice is about one gram, the total weight of rice is:

$$W = 2(A + B)^6 \text{ gr} = 2(A + B)^{(2 + 2 + 2)} \text{ gr} = 2(A + B)^{2 \times 3} \text{ gr}$$

Now we must discuss about the expansion of the binomial $2(A + B)^6$. In the Middle Ages, how did the old man expand the binomial above? As we know, between 540 and 250 BC, Pythagoras was the first one who discovered the theory of irrationals. Based on this theory, the Pythagoreans, including Euclid, discovered the identities:

$$(A + B)^2 = A^2 + 2AB + B^2 \text{ and } (A - B)^2 = A^2 - 2AB + B^2$$

ANCIENT AND MODERN MATHEMATICS

The fall of Alexandria in 391 AD and the event of Hypatia in 415 AD caused a great exodus from Alexandria of many Greek mathematicians and scientists who settled in Arab countries, mostly in Persia and some in India. Eventually, the Persians and Hindus acquired the knowledge of mathematics and sciences from the Greek mathematicians and scientists.

In 1303, a Chinese mathematician named Chu Shih-Chieh discovered a triangle to determine the coefficients in the binomial expansions. This triangle is older than that of Pascal (1665), about 360 years to be exact. To be able to discover it, of course he had to know the particular binomial expansions that might exist from his time up to the previous centuries in China as following:

$(A + B)^0 \Rightarrow \qquad\qquad\qquad 1$

$(A + B)^1 \Rightarrow \qquad\qquad\quad 1 \qquad 1$

$(A + B)^2 \Rightarrow \qquad\quad 1 \qquad 2 \qquad 1$

$(A + B)^3 \Rightarrow \quad 1 \qquad 3 \qquad 3 \qquad 1$

Therefore, if that old man who lived in Persia or India might expand the binomial $2(A + B)^6$ by $2(A + B)^{(2 + 2 + 2)} =$

$2(A + B)^2 \times (A + B)^2 \times (A + B)^2$ or by $2(A + B)^{2.3} = 2[(A + B)^2]^3$ as following:

A. He might calculate the weight by expanding the product:

$$W = 2(A + B)^2 \times (A + B)^2 \times (A + B)^2 = 2(A^2 + 2AB + B^2) \times (A^2 + 2AB + B^2) \times (A^2 + 2AB + B^2)$$

If we substitute $A = 10^3$ and $B = 24$, we have:

$= 2(10^6 + 48.10^3 + 576) \times (10^6 + 48.10^3 + 576) \times (10^6 + 48.10^3 + 576) = 2(1 + 0.048 + 0.000576) \times 10^6 \times (1 + 0.048 + 0.000576) \times 10^6 \times (1 + 0.048 + 0.000576) \times 10^6$ (1)

Because the problem only needs the accuracy of five significant digits, we can round 0.000576 up to 0.0006. Therefore, we can rewrite (1) as:

$W = 2[(1.0486) \times 10^6] \times [(1.0486) \times 10^6] \times [(1.0486) \times 10^6] = 2(1.0486)^2 \times (1.0486) \times 10^{18} = 2[1 + 2 \times 0.0486 + (0.0486)^2] \times (1.0486) \times 10^{18}$

Where:

$W = 2(1 + 0.0972 + 0.00236) \times (1.0486) \times 10^{18}$ (2)

Because we just round 0.000576 up to 0.0006 by adding 0.000024 and the amount 0.0972 by adding 0.000048, we can round 0.00236 to 0.0023 by minus 0.00006. Thus, we can rewrite (2) as below:

$$W = 2(1.0995) \times (1.0486) \times 10^{18} = (2.3058) \times 10^{18}.$$

But 10^6 gr = 1 ton.

Therefore, that old man got W = 2.3058 trillions tons.

B. He might expand:

$$W = 2[(A + B)^2]^3 = 2(A^2 + 2AB + B^2)^3$$

If we substitute $A = 10^3$ and $B = 24$, we have:

$$W = 2(10^6 + 48.10^3 + 576)^3 = 2[(1.0486).10^6]^3 = 2(1.0486)^3 \times 10^{18},$$
or $W = 2(1 + 0.0486)^3 \times 10^{18} = 2[1 + 3 \times 0.0486 + 3(0.0486)^2 + \cancel{(0.0486)^3}] \times 10^{18}$ (3).

Because we already rounded the number 3 x 0.0486 up by adding 3 x 0.000024, we can ignore $(0.0486)^3$ in (3). Thus, the old man also got W = 2.3058 trillions tons.

C. If he lived in China, he might also use the two methods above. Moreover, he might expand the binomial $2(A + B)^6$ by using the triangle of Chu Shil-chieh, if he were a contemporary of that Chinese mathematician. This triangle is as follows:

```
              1
           1     1
        1     2     1
     1     3     3     1
  1     4     6     4     1
1    5    10   10    5    1
1  6    15   20   15    6    1
-  -   -    -    -    -   -
```

$W = 2(A + B)^6 = 2(A^6 + 6A^5B + 15A^4B^2 + 20A^3B^3 + 15A^2B^4 + 6AB^5 + B^6)$.

If we substitute $A = 10^3$ and $B = 24$, we have:

$W = 2[10^{18} + 6 \times 10^{15} \times 24 + 15 \times 10^{12} \times (24)^2 + 20 \times 10^9 \times (24)^3 + 15 \times 10^6 \times (24)^4 + 6.10^3 \times (24)^5 + (24)^6] = 2[1 + 6 \times 0.024 + 15 \times (0.024)^2 + 20(0.024)^3 + \cancel{15(0.024)^4 + 6(0.024)^5 + (0.024)^6}] \times 10^{18}$.

Because the problem allows us to calculate the amounts of rice with an accuracy of five significant digits, we can ignore all the terms from $15(0.024)^4$ to $(0.024)^6$ in the relationship above. We can write:

$W = 2[1 + 6 \times 0.024 + 15(0.024)^2 + 20(0.024)^3] \times 10^{18} = 2[1 + 0.14400 + 0.00864 + 0.00028) \times 10^{18} = 2.3058.10^{18}$ gr $= 2.3058.10^{12}$ tons $= 2.3058$ trillion tons.

The royal mathematicians made this result overnight. The next morning, it was reported to the king. Looking at this report, his smiling face was shut down perpetually because it showed the weight of rice that the king must pay to the old man:

W = 2.3058 trillion tons.

Note: To produce such that amount of rice, the whole world today must take at least two thousand years. Even humans in the meantime had high technology products on hand! Hence, that old man was indeed a genius mathematician!

Chapter 2

Problem 3: The Hundred Fowls (I)

A man named Zhang Qiujian made this problem in China about 450 AD. A farmer has one hundred fowls that cost one hundred Chinese piasters. The fowls consist of cocks, hens, and chicken. Each cock costs five piasters, each hen costs three piasters, and each chicken costs one-third piaster.

Question 1	Question 2	Question 3
How many cocks are there?	How many hens are there?	How many chickens are there?

Solution

Set x as the number of cocks, y as the number of hens, and z as the number of chickens. According to the problem, we have the two following equations:

$$x + y + z = 100 \quad (1)$$

and

$$5x + 3y + \frac{z}{3} = 100 \quad (2)$$

The equation (1) gives us:

$$z = 100 - x - y$$

If we substitute z by (100 − x − y) in the equation (2), we have:

$$5x + 3y + \frac{100 - x - y}{3} = 100$$

If we set 3 as the common denominator of the first member of the equation, we get:

$$\frac{15x + 9y + 100 - x - y}{3} = 100$$

Consequently, we deduce:

$$14x + 8y + 100 = 300$$

If we rearrange and simplify, the equation becomes:

$$7x + 4y = 100 \quad (3)$$

From the equation (3), we have:

$$y = \frac{100 - 7x}{4} = 25 - x - \frac{3x}{4} \quad (4) \text{ and}$$

$$x = \frac{100 - 4y}{7} = 14 - y + \frac{3y + 2}{7} \quad (5)$$

Now we must choose among (4) and (5). Which one would make our calculation to be easier? We recognize that y and x must be the integers. Thus the term $\frac{3x}{4}$ and $\frac{3y+2}{7}$ in (4)

and (5) also must be the integers. But $\dfrac{3x}{4}$ is more simple than $\dfrac{3y+2}{7}$, so we find x in $\dfrac{3x}{4}$ much easier than y in $\dfrac{3y+2}{7}$.
Therefore, we choose the relation (4): $y = 25 - x - \dfrac{3x}{4}$.

In this relation, if y is the integer, x in $\dfrac{3x}{4}$ must be the integer. Therefore, we deduce:

$x_1 = 4, x_2 = 8$

and

$x_3 = 12$

If we substitute $x_1 = 4$, $x_2 = 8$, and $x_3 = 12$ in (4), we have:

$y_1 = 18$, $y_2 = 11$, and $y_3 = 4$

If we substitute x_1 and y_1 in equation (1), we get:

$z_1 = 78$

If we substitute x_2 and y_2 in equation (1), we get:

$z_2 = 81$

If we substitute x_3 and y_3 in (1), we get:

$z_3 = 84$

So to summarize the three answers:

$x_1 = 4$	$y_1 = 18$	$z_1 = 78$
$x_2 = 8$	$y_2 = 11$	$z_2 = 81$
$x_3 = 12$	$y_3 = 4$	$z_3 = 84$

ANCIENT AND MODERN MATHEMATICS

Note: This is the original problem; the author made the similar problems in the next pages.

Problem 4: The Hundred Fowls (2)

A farmer has one hundred fowls, of which the total weights are one hundred kilograms. The fowls consist of cocks, hens, and chickens. The weight of each cock, each hen, and each chick would be three kilograms, two kilograms, and fifty grams respectively.

Question 1	Question 2	Question 3
How many cocks are in 100 fowls above?	How many hens are in 100 fowls above?	How many chickens are in 100 fowls above?

Solution

Set x as the number of cocks, y as the number of hens, and z as the number of chickens. According to the hypothesis, we have two following equations:

$$x + y + z = 100 \quad (1)$$

and

$$3x + 2y + \frac{z}{20} = 100 \quad (2)$$

Equation (1) gives us:

$$z = 100 - x - y$$

If we substitute z by (100 − x − y) in (2), we have:

$$3x + 2y + \frac{100 - x - y}{20} = 100$$

If we set 20 as the common denominator of the first member on the lefthand side of the equation, we have:

$$\frac{60x + 40y + 100 - x - y}{20} = 100$$

Consequently, we deduce:

$$59x + 39y = 1,900$$

Therefore, we have:

$$y = \frac{1,900 - 59x}{39} = 49 - \frac{11}{39} - x - \frac{20x}{39}$$

or

$$y = 49 - x - \frac{20x + 11}{39} \quad (3)$$

Because y is a positive integer, so the term $\frac{20x+1}{39}$ in (3) must be an integer Q < 49 − x. We can write:

$$\frac{20x + 11}{39} = Q$$

or

20x+11=39Q

Therefore, we have:

$$x = \frac{39Q-11}{20} = \frac{40Q-Q-11}{20}$$

or

$$x = 2Q - \frac{Q+11}{20} \quad (4)$$

Because x is a positive integer, the term $\frac{Q+11}{20}$ must be an integer. Therefore: Q = 9, 29, 49 ... We only take Q = 9. If we take Q = 29 or 49, we will get y < 0. This is impossible.

If we substitute Q = 9 in equation (4), we have:

x = 17

If we substitute x = 17 in equation (3), we have:

y = 23

If we substitute x = 17

and

y = 23 in equation (1), we have:

z = 60

Note: This problem has only one answer, and it is made by the author.

Problem 5: The Thousand Fowls

A farmer has one thousand fowls, of which the total weights are one thousand kilograms. The fowls consist of cocks, hens, and chickens. The weight of each cock, each hen, and each chick would be 3 kilograms, 2.5 kilograms, and 0.11 kilograms respectively.

Question 1	Question 2	Question 3
How many cocks are in 1,000 fowls above?	How many hens are in 1,000 fowls above?	How many chicken are in 1,000 fowls above?

Solution

Set x as the number of cocks, y as the number of hens, and z as the number of chickens. According to the hypothesis, we have two following equations:

$$x + y + z = 1{,}000 \quad (1)$$

and

$$3x + 2.5y + 0.11z = 1{,}000 \quad (2)$$

Equation (1) gives us:

$$z = 1{,}000 - x - y$$

If we substitute z by (1,000 − x − y) in (2), we have:

3x + 2.5y + 0.11 (1,000 − x − y) = 1,000

If we multiply two members of this equation by 100, we have:

300x + 250y + 11 (1,000 − x − y) = 100,000

If rearrange, we obtain:

289x + 239y = 89,000 (3)

Question I

From the equation (3), we obtain:

$$y = \frac{89,000 - 289x}{239}$$

Because y is a positive integer, therefore, the fraction above must also be a positive integer. That means x must be such a positive integer that satisfies the numerator (89,000 − 289x) to be divided evenly by the denominator 239. But the numerator and denominator of that fraction contain the large numbers; eventually, it is difficult to find x. Therefore, we use the iterative methods to determine x as following:

A. $y = \dfrac{89{,}000 - 289x}{239}$

or

$$= 372 + \dfrac{92}{239} - x - \dfrac{50x}{239}$$

$$= 372 - x - \dfrac{50x - 92}{239}$$

or

$$y = 372 - x - \dfrac{2(25x - 46)}{239} \quad (4)$$

B. We already know that y is an integer and the terms 372 and $-x$ in (4) are the integers, so the term $\dfrac{2(25x - 46)}{239}$ or $\dfrac{25x - 46}{239}$ also must be an integer. Therefore, we can write:

$$\dfrac{25x - 46}{239} = Q_1 \quad (Q_1 \text{ is an integer}).$$

From this fraction, we obtain:

$$x = \dfrac{239 Q_1 + 46}{25}$$

$$= 10 Q_1 - \dfrac{11 Q_1}{25} + 2 - \dfrac{4}{25}$$

or

$$x = 10 Q_1 + 2 - \dfrac{11 Q_1 + 4}{25} \quad (5)$$

C. Once again x, $10Q_1$, and 2 in (5) are integers, so the term $\dfrac{11Q_1+4}{25}$ must be integer. Therefore, we again can write:

$$\dfrac{11Q_1+4}{25} = Q_2 \quad (Q_2 \text{ is an integer})$$

Thus, we deduce:

$$Q_1 = \dfrac{25Q_2-4}{11}$$

$$= 2Q_2 + \dfrac{3Q_2}{11} - \dfrac{4}{11}$$

or

$$Q_1 = 2Q_2 + \dfrac{3Q_2-4}{11} \quad (6)$$

D. And again Q_1 and $2Q_2$ in (6) are the integers, so the term $\dfrac{3Q_2-4}{11}$ must be an integer. So we again can write:

$$\dfrac{3Q_2-4}{11} = Q_3 \quad (Q_3 \text{ is an integer})$$

Then we obtain:

$$Q_2 = \dfrac{11Q_3+4}{3}$$

$$= 4Q_3 - \dfrac{Q_3}{3} + 1 + \dfrac{1}{3}$$

or

$$Q_2 = 4Q_3 + 1 - \dfrac{Q_3-1}{3} \quad (7)$$

E. From (7), the term $\dfrac{Q_3 - 1}{3}$ must be also an integer. Looking at this fraction, we recognize easily that Q_3 must be 1, 4, 7 ... We only take $Q_3 = 1$. (If $Q_3 = 4$ or 7, it will give us y < 0, which is impossible.) If we substitute $Q_3 = 1$ in (7), we have:

$Q_2 = 5$

If we substitute $Q_2 = 5$ in (6), we have:

$Q_1 = 11$

If we substitute $Q_1 = 11$ in (5), we have:

x = 107

If we substitute x = 107 in (4), we have:

y = 243

If we substitute x = 107 and y = 243 in (1), we obtain:

z = 650

Question 2

In the first option, we choose y as the function of x to determine x. In this second option, we choose x as the function of y to determine y. Therefore, from equation (3), we have:

$$x = \dfrac{89,000 - 239y}{289}$$

Because x is a positive integer, so the fraction above also must be a positive integer. That means y must be such a positive integer that satisfies the numerator (89,000 − 239y) to be divided evenly by denominator 289. However, the numerator and denominator of that fraction contain the large numbers. Eventually, it is very hard to determine y. Therefore, we use the iterative methods to find y as below:

A. $$x = \frac{89,000 - 239y}{289} = 308 - \frac{12}{289} - y + \frac{50y}{289}$$

or

$$x = 308 - y + \frac{2(25y - 6)}{289} \quad (4)$$

B. Because x is an integer and the terms 308 and −y in (4) are the integers, thus the term $\frac{2(25y-6)}{289}$ or $\frac{25y-6}{289}$ also must be an integer. Therefore, we can write:

$$\frac{25y - 6}{289} = Q_1 \text{ (where } Q_1 \text{ is an integer)}$$

This fraction gives us:

$$y = \frac{289Q_1 + 6}{25}$$

$$= 12Q_1 - \frac{11Q_1}{25} + \frac{6}{25}$$

or

$$y = 12Q_1 - \frac{11Q_1 - 6}{25} \quad (5)$$

C. Again, y and $12Q_1$ in (5) are integers, so the term $\frac{11Q_1 - 6}{25}$ must be an integer. Therefore, we can write:

$$\frac{11Q_1 - 6}{25} = Q_2 \text{ (where } Q_2 \text{ is an integer)}$$

Hence, we deduce:

$$Q_1 = \frac{25Q_2 + 6}{11}$$

or

$$Q_1 = 2Q_2 + \frac{3Q_2 + 6}{11} \quad (6)$$

D. Once again Q_1 and $2Q_2$ in (6) are the integers, so the term $\frac{3Q_2 + 6}{11}$ also must be an integer. And again, we can write:

$$\frac{3Q_2 + 6}{11} = Q_3 \text{ (where } Q_3 \text{ is an integer)}$$

Thus, we obtain:

$$Q_2 = \frac{11Q_3 - 6}{3}$$

or

$$Q_2 = 4Q_3 - 2 - \frac{Q_3}{3} \quad (7)$$

But Q_2, $4Q_3$, and -2 in (7) are the integers, so $\frac{Q_3}{3}$ also must be an integer. We recognize easily $Q_3 = 3, 6, 9 \ldots$ We only accept:

$Q_3 = 3$ (if $Q_3 = 6, 9 \ldots$ it leads to $x < 0$)

If we substitute $Q_3 = 3$ in (7), we have:

$Q_2 = 9$

If we substitute $Q_2 = 9$ in (6), we have:

$Q_1 = 21$

If we substitute $Q_1 = 21$ in (5), we have:

$y = 243$

If we substitute $y = 243$ in (4), we have:

$x = 107$

If we substitute $y = 243$ and $x = 107$ in (1), we have:

$z = 650$.

Question 3

In this option, we choose x as the function of z to find z. To do so, we first deduce y from the equation (1), as below:

$y = 1{,}000 - x - z$

If we substitute y = 1,000 − x − z in the equation (2) and then multiply its two members by 100, we obtain:

$$300x + 250(1,000 - x - z) + 11z = 100,000$$

Or we can rewrite:

$$50x - 239z + 250,000 = 100,000$$

Thus, we deduce:

$$x = \frac{239z - 150,000}{50} > 0$$

or

$$x = \frac{239z}{50} - 3,000 > 0 \quad (4)$$

Because x and −3,000 in (4) are the integers, so $\frac{239z}{50}$ must be an integer. That means z must be such a positive integer that is divided evenly by 50. From (4), we deduce: $239\left(\frac{z}{50}\right) > 3,000$

We can rearrange:

$$\frac{z}{50} > \frac{3,000}{239} = 12 + \frac{132}{239}$$

Thus, we deduce:

$$\frac{z}{50} = 13, \ 14, \ 15, \dots..$$

We only accept $\frac{z}{50}=13$ because, if $\frac{z}{50}=14\ldots$, it gives us z = 700 and x = 346; therefore: z + x = 1,046

This is impossible. Therefore, we get:

z = 650

If we substitute z = 650 in (4), we get:

x = 107

If we substitute z = 650 and x = 107 in equation (1), we get:

x = 243

Chapter 3

Problem 6: The Problem of Sunzi

Sunzi, a man of the Han dynasty of China in about 250 C.E., made this problem. In ancient times, his position would be considered as general chief of staff of the Chinese armed forces. He frequently went to the front line to inspect his troop units with such a manner:

1. He ordered his troops to form a neatly arranged formation. If we look at the formation from the front to the rear, it consists of three columns. If we look from the side, it has many rows. Each row consists of three soldiers, and of course, the last row is not a complete one but has only a soldiers ($3 > a \geq 0$).
2. He ordered this unit to form a formation of five columns and many rows. Each row has five soldiers, and the last row has only b soldiers ($5 > b \geq 0$).
3. The formation consists of seven columns and many rows. Each row has seven soldiers, and the last row has only c soldiers ($7 > c \geq 0$).

All these three steps above must take their turn to parade in front of SUNZI. Then he only counted the numbers of soldiers in the last row, and he could know the present soldiers of that unit right away. For example, he recognized two soldiers in the last row of the first parade, he got three soldiers in the last row of the second parade, and he got two soldiers in the last row of the third parade. How many present soldiers are in that unit?

Solution

There are two solutions: the original solution of Sunzi and the improved solution of the author.

Solution I: Sunzi's Original Solution

This solution is found in the ancient Chinese mathematics on the website, as below. Set N as the numbers of the present soldiers we have:

$N = 140 + 63 + 30 - 210 = 23$ (the least numbers)

Comment: This solution is not clear because it does not prove the existence of the numbers 140, 63, 30, and 210 in the relationship of N above.

I think this is a practical rule, or those numbers might come from two formulas below:

$N = 70a + 21b + 15c - 210$ (if $70a + 21b + 15c > 210$)

or

$N = 70a + 21b + 15c - 105$ (if $70a + 21b + 15c < 210$)

These formulas give us the least N. After that, if we want to seek the larger number N, we add 105 to N as many times as to reach this number.

In case of the big units, such as battalion, regiment, division, or corps, this solution cannot guess the exact numbers N because the difference of the consecutive N_1, N_2, N_3 ... are only 105.

Solution 2: Improved Solution of the Author

Here is the improved solution. First, we find the least common multiple of 3, 5, and 7:

$3 \times 5 \times 7 = 105$

Second, we set N as the numbers of the present soldiers of the unit. Then we have the total complete rows of the first formation:

$$\frac{N-a}{3} = K_a$$

We have the total complete rows of the second formation:

$$\frac{N-b}{5} = K_b$$

We have the total complete rows of the third formation:

$$\frac{N-c}{7} = K_c$$

We recognize: $K_a > K_b \geq K_c$

Now we establish the relationship as:

$$\frac{N-c}{7} + \frac{N-b}{5} - \frac{N-a}{3} = K_c + K_b - K_a = K \, (K = -1, 0, 1, 2, 3 \ldots)$$

Let 105 be the common denominator of the first member of that relationship we have:

$$\frac{15(N-c) + 21(N-b) - 35(N-a)}{105} = K \quad (K = -1, 0, 1, 2, 3 \ldots)$$

Hence, we deduce:

N − 15c − 21b + 35a = 105K

Finally, we have the formula:

N = 105K + 15c + 21b − 35a (K = −1, 0, 1, 2, 3 …)

Application

1. In the example of the original solution, we recognize $a = 2$, $b = 3$, and $c = 2$. So we can apply the formula above. Then we obtain the least N with $K = 0$: $N_1 = 105 \times 0 + 15 \times 2 + 21 \times 3 - 35 \times 2 = 30 + 63 - 70 = 23$ (a platoon size). With $K = 1$ we have: $N_2 = 105 + 30 + 63 - 70 = 128$ (a company size).

2. In case of $a = 0$, $b = 4$, and $c = 6$, we have:
   $$N = 105K + 90 + 84 - 0 = 105K + 174$$
   So the least N with $K = -1$:
   $$N = -105 + 174 = 69$$

3. In case of $a = 1$, $b = 1$, and $c = 1$, we have:
   $N_1 = 15 + 21 - 35 = 1$ (the least N with $K = 0$)
   and
   $N_2 = 105 + 1 = 106$ (more practical than $N_1 = 1$)

4. In case of $a = 2$, $b = 4$, and $c = 6$, we have:
   $$N = 105K + 90 + 84 - 70 = 105K + 104$$
   so
   $N = 104$ (with $K = 0$)

5. In case of $a = 2$, $b = 0$, and $c = 0$, we have the least N as:
   $$N = 105K + 0 + 0 - 70 = 105 - 70$$
   so
   $N = 35$ (with $K = 1$)

The original problem of Sunzi cannot answer the exact numbers N soldiers in a big unit such as a regiment, division, or corps because it gives us too many N with the difference of 105. Therefore, I think of the similar problems that can be solved to find the least N in any big units of troops in the following pages.

Problem 7: The Improved Problem of SUNZI to Inspect Battalion

Consider a unit of troops within the battalion size. First, this unit arranges in a formation that consists of seven columns (looked from the front to the rear) and a certain number of rows (looked from the side of the formation). Each row has seven soldiers; the last row has only a ($0 \leq a < 7$) soldiers. Second, this unit arranges in another formation with nine columns and a certain rows. Each row has nine soldiers; the last has b ($0 \leq b < 9$) soldiers. Third, this unit once again makes another formation with eleven columns and a certain rows. Each row has eleven soldiers; the last has only c ($0 \leq c < 11$) soldiers. Find the formula to know the present soldiers of that unit.

Solution

First, we calculate the least common multiple of 7, 9, and 11:

$$7 \times 9 \times 11 = 693$$

Second, we set N as the numbers of soldiers in this unit. Then we have the total complete rows of the first formation:

$$\frac{N-a}{7} = K_a$$

We have the total complete rows of the second formation:

$$\frac{N-b}{9} = K_b$$

We have the total complete rows of the third formation:

$$\frac{N-c}{11} = K_c$$

We recognize:

$$K_a \geq K_b \geq K_c$$

Now we establish the relationship as following:

A. Set 693 as the common denominator of $\frac{N-a}{7} = K_a$, $\frac{N-b}{9} = K_b$, and $\frac{N-c}{11} = K_c$. We obtain:

$$\frac{99(N-a)}{693} = K_a, \quad \frac{77(N-b)}{693} = K_b, \quad \frac{63(N-c)}{693} = K_c$$

Look at these three numbers if we perform the calculation such as:

$K_a + K_b + K_c = K_1$

$K_x + K_b + K_a = K_2$

$m.K_a + n.K_b + p.K_c = K_3$

$n.K_b + p.K_c - m.K_a = K_4$

There are so many formulas to find N in connection with a, b, c, and K. Remember a, b, and c are given and K is variable. But not all of those formulas can we use to find N easily. For example:

$K_a + K_b + K_c = K$

$$\frac{99(N-a)}{693} + \frac{77(N-b)}{693} + \frac{63(N-c)}{693} = K$$

If we rearrange and simplify, we obtain:

239N = 693K + 99a + 77b + 63c

The relationship above shows that 239 is the coefficient of N; it is a large number; thus, it causes the calculation to be very difficult to find N.

B. So we try to find out such a way that gives us the least coefficient of N to easily solve the problem. To do so, we consider 99N of K_a, 77N of K_b, and 63N of K_c. Then we establish two relationships by subtracting as below:

99N − 77N = 22N (1)

and

77N − 63N = 14N (2)

If we multiply (1) by 2 and multiply (2) by 3 we have:

2(99N − 77N) = 44N (3)

and

3(77N − 63N) = 42N (4)

If we then subtract (3) by (4), we have:

2(99N − 77N) − 3(77N − 63N) = 2N

We can rewrite:

2 x 99N − 5 x 77N + 3 x 63N = 2N (5)

Thus, we get 2 as the least coefficient of N. Based on the relationship (5), we can establish the formula between N and a, b, c, and K as following:

$$\frac{2\times 99(N-a)}{693} - \frac{5\times 77(N-b)}{693} + \frac{3\times 63(N-c)}{693} = 2K_a - 5K_b + 3K_c = K$$

Thus, we deduce:

$$\frac{2N - 198a + 385b - 189c}{693} = K$$

If we rearrange, we obtain:

$$N = \frac{693k + 198a - 385b + 189c}{2}$$

Or we can rewrite:

$$N = \frac{692K + K + 198a - 386b + b + 188c + c}{2}$$

Finally, we get the formula:

$$N = 346k + 99a - 193b + 94c + \frac{k+b+c}{2} \quad (-4 \leq k \leq 6)$$

Set $2Q + R = b + c$ (Q and R known eventually)

$0 \leq Q \leq 9$ and $0 \leq R \leq 1$

$$N = 346k + 99a - 193b + 94c + Q + \frac{K+R}{2} \quad (-4 \leq k \leq 6)$$

Remarks

- Because N is the integer and the terms 346K, 99a, 193b, 94c, and Q are also the integers, the term $\frac{K+R}{2}$ must be an integer. That means K must be such a number that satisfies $\frac{K+R}{2}$ to be divisible evenly by 2.
- In this problem, we have the least N to be $0 < N \leq 693$. To know the least N, we must find the least K, which satisfies the condition $0 < N \leq 693$.

- In the formula:

$$N = 346k + 99a - 193b + 94c + \frac{k+b+c}{2}$$

To know Q, we take (b + c) to be divided by 2. Then we get a quotient Q and a remainder R. The formula becomes:

$$N = 346K + 99a - 193b + 94c + Q + \frac{K+R}{2}$$

Whence we know R, we can get K, $\frac{K+R}{2}$, and 346K in the table of data. Eventually, we know the least N.

- The following table contains the data of 346K, 99a, −193b, 94c, R, K, and $\frac{K+R}{2}$. So we can use it to find the least N with some minor calculations, and we get the answer in about one minute.

- The table of data for the problem 7 with the formula:

$$N = 346K + 99a - 193b + 94c + Q + \frac{K+R}{2}$$

2Q+R = b + c

Ancient and Modern Mathematics

a	99a		b	−193b		c	94c		R	K	K+R / 2	346K
1	99		1	−193		1	94		0	−4	−2	−1,384
2	198		2	−386		2	188		0	−2	−1	−692
3	297		3	−579		3	282		0	0	0	0
4	396		4	−772		4	376		0	2	1	692
5	495		5	−965		5	470		0	4	2	1,384
6	594		6	−1,158		6	564		0	6	3	2,076
			7	−1,351		7	658		1	−3	−1	−1,038
			8	−1,544		8	752		1	−1	0	−346
						9	846		1	1	1	346
						10	940		1	3	2	1,038
									1	5	3	1,730

Exercises

Question I

Find the least N so that:

- Upon division by 7, we get a remainder of 6
- Upon division by 9, we get a remainder of 0
- Upon division by 11 we get a remainder of 10

Solution

Consider the formula:

$$N = 346K + 99a - 193b + 94c + Q + \frac{K+R}{2}$$

Because a = 6, b = 0, c = 10, and 2Q + R = 10, Q = 5 and R = 0. The formulas becomes:

$$N = 346K + 594 - 0 + 940 + 5 + \frac{K+0}{2}$$

$$= 346K + 1,539 + \frac{K}{2}$$

In the table, the correspondences of R = 0 are K = −4, $\frac{K+R}{2} = -2$, and 346K = −1,384. Therefore, we obtain:

N = −1,384 + 1,539 − 2. Or N = 153

Question 2

Find the least N so that:
- Upon division by 7, we get a remainder of 0
- Upon division by 9, we get a remainder of 8
- Upon division by 11, we get a remainder of 0

Solution

In the formula:

$$N = 346K + 99a - 193b + 94c + Q + \frac{K+R}{2}, \text{ we have:}$$

a = 0, b = 8, c = 0, and 2Q + R = 8 + 0

so

Q = 4 and R = 0

We apply the formula we have above:

$$N = 346K + 0 - 1{,}544 + 0 + 4 + \frac{K+0}{2}$$

In this case, the correspondences of R = 0 are K = 6, $\frac{K+R}{2} = 3$, and 346K = 2,076. Therefore, we obtain:

$$N = 2{,}076 - 1{,}544 + 4 + 3. \text{ Or } N = 539$$

Question 3

Find the least N so that:

- Upon division by 7, we get a remainder of 1
- Upon division by 9, we get a remainder of 1
- Upon division by 11, we get a remainder of 1

Solution

In the formula:

$$N = 346K + 99a - 193b + 94c + Q + \frac{K+R}{2}$$

a = 1, b = 1, c = 1, and 2Q + R = 2

so

Q = 1 and R = 0

If we substitute a, b, c, Q by 1, and R by 0 in the formula above, we have:

$$N = 346K + 99 - 193 + 94 + 1 + \frac{K+0}{2}$$

$$N = 346K + 1 + \frac{K+0}{2}$$

In the table, the correspondences of R = 0 are K = 0 or K = 2, $\frac{K+R}{2} = 0$ or $\frac{K+R}{2} = 1$, and 346K = 0 or 346K = 692.

If K = 0, we have:

N = 1 (not practical)

If K = 2, we have:

N = 694

But N = 694 is more practical than N = 1. Therefore, we get the answer:

N = 694

Question 4

Find the least N so that:
- Upon division by 7, we get a remainder of 6
- Upon division by 9, we get a remainder of 8
- Upon division by 11, we get a remainder of 10

Solution

We already recognize the formula:
$$N = 346K + 99a - 193b + 94c + Q + \frac{K+R}{2}$$

By the hypothesis of the problem, we have:
   a = 6, b = 8, c = 10, and 2Q + R = 18

So:
   Q = 9 and R = 0

Therefore, we have:
$$N = 346K + 594 - 1{,}544 + 940 + 9 + \frac{K+0}{2}$$

Correspondent to R = 0, we get K = 2, $\frac{K+R}{2} = 1$, and 346K = 692. Hence, we obtain:
   N = 692

Question 5

Find the least N so that:
- Upon division by 7, we get a remainder of 3
- Upon division by 9, we get a remainder of 7
- Upon division by 11, we get a remainder of 4

Solution

First, we write down the formula:

$$N = 346K + 99a - 193b + 94c + Q + \frac{K+R}{2}$$

According to the hypothesis, we have:

$a = 3, b = 7, c = 4$, and $2Q + R = 7 + 4 = 11$

So:

$Q = 5$ and $R = 1$

The table gives us the first step:

$$N = 346K + 297 - 1{,}351 + 376 + 5 + \frac{K+1}{2}$$

$$= 346K - 673 + \frac{K+1}{2}$$

The table gives us the second step. The correspondences of $R = 1$ are $K = 3$, $\frac{K+1}{2} = 2$, and $346K = 1{,}038$. Thus, we obtain:

$N = 1{,}038 - 673 + 2$

or

$N = 367$

Problem 8: The Improved Problem of SUNZI to Inspect Regiment

Consider a unit of troops with the size between a squad to a regiment. First, this unit arranges in a formation that consists of thirteen columns and many rows. Each row has thirteen soldiers; the last row has only a ($0 \leq a < 13$) soldiers. Second, this unit arranges in another formation that consists of seventeen columns and many rows. Each row has seventeen soldiers. The last row has only b ($0 \leq b < 17$) soldiers. Third, this unit once again arranges in a formation that consists of nineteen columns and many rows. Each row has nineteen soldiers. The last row has only c ($0 \leq c < 19$) soldiers. How many soldiers are in this unit?

Solution

We first find the least common multiple of 13, 17, and 19: 13 x 17 x 19 = 4,199. Second, we set N as the present soldiers in this unit. Then we have the total complete rows of the first formation:

$$\frac{N-a}{13} = K_a$$

We have the total complete rows of the second formation:

$$\frac{N-b}{17} = K_b$$

We have the total complete rows of the third formation:

$$\frac{N-c}{19} = K_c$$

We recognize: $K_a \geq K_b \geq K_c$. Now we establish the relationship as follows:

A. Set 4,199 as the common denominator of $\frac{N-a}{13} = K_a$, $\frac{N-b}{17} = K_b$, and $\frac{N-c}{19} = K_c$. We have:

$$\frac{323(N-a)}{4,199} = K_a, \quad \frac{247(N-b)}{4,199} = K_b, \quad \frac{221(N-c)}{4,199} = K_c$$

B. Like problem 6, we find the minimum coefficient of N. To do so, we consider 323N of K_a, 247N of K_b, and 221N of K_c. Then we make two relationship by subtracting as below:

323N − 247N = 76N (1)

and

247N − 221N = 26N (2)

If we multiply (2) by 3, we have:

3(247N − 221N) = 78 N (3)

If we subtract (3) by (1), we have:

$$3(247N - 221N) - (323N - 247N) = 78N - 76N = 2N$$

Or we can write:

$$4 \times 247N - 3 \times 221N - 323N = 2N \quad (4)$$

Thus we get 2 as the minimum coefficient of N. Based on the relationship (4), we can establish the relationship between N and a, b, c, and

$$K: \frac{4 \times 247(N-b) - 3 \times 221(N-c) - 323(N-a)}{4{,}199} = 4K_b - 3K_c - K_a = K$$

Hence, we deduce:

$$988(N - b) - 663(N - c) - 323(N - a) = 4{,}199K$$

or

$$2N - 988b + 663c + 323a = 4{,}199K$$

So we have:

$$N = \frac{4{,}199K + 988b - 663c - 323a}{2}$$

Or we can write as:

$$N = \frac{4{,}200K - K + 2 \times 494b - 662c - c - 322a - a}{2}$$

Finally, we have the formulas below:

$N = 2{,}100K + 494b - 331c - 161a - \dfrac{K+a+c}{2}$	$K = -2, -1,$ $0, 1, 2, 3, 4)$
Set: $2Q + R = a + c$ (Q and R known eventually) $0 \leq Q \leq 15$ and $0 \leq R \leq 1$	
$N = 2{,}100K + 494b - 331c - 161a - Q - \dfrac{K+R}{2}$	

Remarks

- Because N is the integer and $2{,}100K$, $494b$, $-331c$, $-161a$, and $-Q$ are also the integers, the term $\dfrac{K+R}{2}$ must be the integer. That means K must be such a number that satisfies $\dfrac{K+R}{2}$ to be divisible by 2.
- In the problem, we recognize $0 \leq N \leq 4{,}199$. This is the least N. To know the least N, must find the least K that must satisfy the condition: $0 \leq N \leq 4{,}199$.

The following is the data for problem 8 with the formula:

$$N = 2{,}100K + 494b - 331c - 161a - Q - \dfrac{K+R}{2}$$

b	494b	c	−331c	a	−161a	R	K	$-\frac{K+R}{2}$	2,100K
1	494	1	−331	1	−161	0	−2	1	−4,200
2	988	2	−662	2	−322	0	0	0	0
3	1,482	3	−993	3	−483	0	2	−1	4,200
4	1,976	4	−1,324	4	−644	0	4	−2	8,400
5	2,470	5	−1,655	5	−805	1	−1	0	−2,100
6	2,964	6	−1,986	6	−966	1	1	−1	2,100
7	3,458	7	−2,317	7	−1,127	1	3	−2	6,300
8	3,952	8	−2,648	8	−1,288	1	5	−3	10,500
9	4,446	9	−2,979	9	−1,449				
10	4,940	10	−3,310	10	−1,610				
11	5,434	11	−3,641	11	−1,771				
12	5,928	12	−3,972	12	−1,932				
13	6,422	13	−4,303						
14	6,916	14	−4,634						
15	7,410	15	−4,965						
16	7,904	16	−5,296						
		17	−5,627						
		18	−5,958						

Exercises

Question I

Find the least N so that:

- Upon division by 13, we get a remainder of 0
- Upon division by 17, we get a remainder of 16
- Upon division by 19, we get a remainder of 0

Solution

Consider the formula:

$$N = 2{,}100K + 494b - 331c - 161a - Q - \frac{K+R}{2}$$

According to the hypothesis, we have:

$a = 0, b = 16,$ and $c = 0$

Because $2Q + R = a + c = 0$, we get:

$Q = 0$ and $R = 0$

The data in the table gives us:

$494b = 7{,}904$

The correspondences of $R = 0$ are $K = -2$ and $-\frac{K+0}{2} = 1$. Thus, the formula above becomes:

$N = 2{,}100 \times (-2) + 7{,}904 + 1 = 3{,}705$

Question 2

Find the least N so that:

- Upon division by 13, we get a remainder of 12
- Upon division by 17, we get a remainder of 0
- Upon division by 19, we get a remainder of 18

Solution

Consider the formula:

$$N = 2{,}100K + 494b - 331c - 161a - Q - \frac{K+R}{2}$$

According to the hypothesis, we have:
$a = 12$, $b = 0$, and $c = 18$

Because $2Q + R = a + c = 30$, so $Q = 15$ and $R = 0$. The data in the table gives us: $494b = 0$, $-331c = -5{,}958$, and $-161a = -1{,}932$. The correspondences of $R = 0$ are $K = 4$ and $-\frac{K+0}{2} = -2$. Therefore, we get the least N below:

$N = 2{,}100 \times 4 - 5{,}958 - 1{,}932 - 15 - 2 = 493$

Question 3

Find the least N so that:
- Upon division by 13, we get a remainder of 11
- Upon division by 17, we get a remainder of 0
- Upon division by 19, we get a remainder of 18

Solution

Consider the formula:

$$N = 2{,}100K + 494b - 331c - 161a - Q - \frac{K+R}{2}$$

We have: a = 11, b = 0, c = 18, and 2Q + R = a + c = 29, so:

Q = 14 and R = 1

The data in the table gives us:

494b = 0, −331c = −5,958, and −161a = −1,771.

The correspondences of R = 1 are K = 5 and $-\frac{K+1}{2} = -3$.
Therefore, the least N is:

N = 2,100 x 5 − 5,958 − 1,771 − 14 − 3 = 2,754

Question 4

Find the least N so that:

- Upon division by 13, we get a remainder of 3
- Upon division by 17, we get a remainder of 9
- Upon division by 19, we get a remainder of 13

Solution

Consider the formula:

$$N = 2{,}100K + 494b - 331c - 161a - Q - \frac{K+R}{2}$$

We have:

$a = 3$, $b = 9$, $c = 13$, and $2Q + R = a + c = 16$

So:

$Q = 8$ and $R = 0$

The data in the table gives us:

$494b = 4{,}446$, $-331c = -4{,}303$, and $-161a = -483$

The correspondences to $R = 0$ is $K=2$ and $-\frac{K+0}{2} = -1$.

Therefore, the least N is:

$N = 2{,}100 \times 2 - 4{,}446 - 4{,}303 - 483 - 8 - 1$

or

$N = 3{,}851$.

Problem 9: The Improved Problem of Sunzi to Inspect Division

Consider a unit of troops within the division size. First, this unit arranges in a formation that consists of twenty-three columns and many rows. Each row has twenty-three soldiers; the last row has only a (0 ≤ a < 23) soldiers. Second, this unit arranges in a formation that consists of twenty-six columns and rows. Each row has twenty-six soldiers; the last row has only b (0 ≤ b < 26) soldiers. Third, this unit once again arranges in a formation that consists of twenty-nine columns and rows; each row has twenty-nine soldiers; the last row has only c (0 ≤ c < 29) soldiers. How many soldiers are in this unit?

Solution

We first find the least common multiple of 23, 26, and 29:

23 x 26 x 29 = 17,342

Second, we set N as the present soldiers in this unit. Then we have the total complete rows of the first formation:

$$\frac{N-a}{23} = K_a$$

We have the total complete rows of the second formation:

$$\frac{N-b}{26} = K_b$$

We have the total complete rows of the third formation:

$$\frac{N-c}{29} = K_c$$

We recognize: $K_a \geq K_b \geq K_c$. Now we establish the relationship as follows:

A. Set 17,342 as the common denominator of $\dfrac{N-a}{23} = K_a$, $\dfrac{N-b}{26} = K_b$, and $\dfrac{N-c}{29} = K_c$. We have:

$$\dfrac{754(N-a)}{17,342} = K_a, \quad \dfrac{667(N-b)}{17,342} = K_b, \quad \dfrac{598(N-c)}{17,342} = K_c$$

B. Like the previous problems, we must find out the least coefficient of N. To do so, we consider 754N of K_a, 667N of K_b, and 598N of K_c. Then we make two relationships by subtracting:

754N − 667N = 87N (1)

and

667N − 598N = 69N (2)

If we multiply (1) by 4 and (2) by 5, we have:

4(754N − 667N) = 4 x 87N = 348N (3)

and

5(667N − 598N) = 5 x 69N = 345N (4)

If we subtract (3) by (4), we have:

4(754N − 667N) − 5(667N − 598N) = 348N − 345N = 3N

Hence, we deduce:

4 x 754N − 9 x 667N + 5 x 598N = 3N (5)

Therefore, we get 3 as the least coefficient of N. Based on the relationship (5), we can establish the formulas between N and a, b, c, and K:

$$K: \frac{4\times 754(N-a) - 9\times 667(N-b) + 5\times 598(N-c)}{17,342} = 4K_a - 9K_b + 5K_c = K$$

Hence, we deduce:

$$4 \times 754(N-a) - 9 \times 667(N-b) + 5 \times 598(N-c) = 17{,}342K$$

Simplifying, we deduce:

$$3N - 3{,}016a + 6{,}003b - 2{,}990c = 17{,}342K$$

Finally, we have:

$$N = \frac{17{,}342K + 3{,}016a - 6{,}003b + 2{,}990c}{3} \quad (K = -6, -5, ..., -1, 0, 1, 2, ..., 8, 9)$$

Or we can write:

$$N = \frac{17{,}343K - K + 3{,}015a + a - 6{,}003b + 2{,}991c - c}{3}$$

Finally, we have the formula:

$$\boxed{\begin{array}{l} N = 5{,}781K + 1{,}005a - 2{,}001b + 997c - \dfrac{K+c-a}{3} \\[4pt] \text{Set } 3Q + R = c - a \ (Q \text{ and } R \text{ known eventually}) \\ -7 \leq Q \leq 9,\ -2 \leq R \leq 2 \\[4pt] N = 5{,}781K + 1{,}005a - 2{,}001b + 997c - Q - \dfrac{K+R}{3} \\[4pt] (K = -6, -5 \ldots -1, 0, 1, 2 \ldots 8, 9) \end{array}}$$

Remarks

To know the least N, we find the least K that satisfies the term $\frac{K+R}{3}$ to be the integer number and also the K must satisfy the condition $0 \leq N \leq 17{,}342$.

The data for the problem 9 with the formula is as follows:

$$N = 5{,}781K + 1{,}005a - 2{,}001b + 997c - Q - \frac{K+R}{3}.$$

a	1,005a	b	−2,001b	c	997c	R	K	$-\dfrac{K+R}{3}$	5,781K
1	1,005	1	−2,001	1	997	−2	−4	2	−23,124
2	2,010	2	−4,002	2	1,994	−2	−1	1	−5,781
3	3,015	3	−6,003	3	2,991	−2	2	0	11,562
4	4,020	4	−8,004	4	3,988	−2	5	−1	28,905
5	5,025	5	−10,005	5	4,985	−2	8	−2	46,248
6	6,030	6	−12,006	6	5,982	−1	−5	2	−28,905
7	7,035	7	−14,007	7	6,979	−1	−2	1	−11,562
8	8,040	8	−16,008	8	7,976	−1	1	0	5,781
9	9,045	9	−18,009	9	8,973	−1	4	−1	23,124
10	10,050	10	−20,010	10	9,970	−1	7	−2	40,467
11	11,055	11	−22,011	11	10,967	0	−6	2	−34,686
12	12,060	12	−24,012	12	11,964	0	−3	1	−17,343
13	13,065	13	−26,013	13	12,961	0	0	0	0
14	14,070	14	−28,014	14	13,958	0	3	−1	17,343
15	15,075	15	−30,015	15	14,955	0	6	−2	34,686
16	16,080	16	−32,016	16	15,952	0	9	−3	52,029
17	17,085	17	−34,017	17	16,949	1	−4	1	−23,124
18	18,090	18	−36,018	18	17,946	1	−1	0	−5,781
19	19,095	19	−38,019	19	18,943	1	2	−1	11,562
20	20,100	20	−40,020	20	19,940	1	5	−2	28,905
21	21,105	21	−42,021	21	20,937	1	8	−3	46,248
22	22,110	22	−44,022	22	21,934	2	−5	1	−28,905
		23	−46,023	23	22,931	2	−2	0	−11,562
		24	−48,024	24	23,928	2	1	−1	5,781
		25	−50,025	25	24,925	2	4	−2	23,124
				26	25,922	2	7	−3	40,467
				27	26,919				
				28	27,916				

Exercises

Question 1

Find the least N so that:

- Upon division by 23, we get a remainder of 22
- Upon division by 26, we get a remainder of 0
- Upon division by 29, we get a remainder of 28

Solution

We first write down the formula:

$$N = 5,781K + 1,005a - 2,001b + 997c - Q - \frac{K+R}{3}$$

Because $a = 22$, $b = 0$, $c = 28$, and $3Q + R = c - a = 6$, we have:

$Q = 2$ and $R = 0$

The data first gives us:

$$N = 5,781K + 22,110 - 0 + 27,916 - 2 - \frac{K+0}{3}$$

$$= 5,781K + 50,024 - \frac{K+0}{3}$$

Second, the data shows $R = 0$ corresponding to $K = -6$ and $5,781K = -34,686$. Thus we have:

$$N = -34,686 + 50,024 + 2 = 15,340$$

Question 2

Find the least N so that:

- Upon division by 23, we get a remainder of 0
- Upon division by 26, we get a remainder of 25
- Upon division by 29, we get a remainder of 0

Solution

We first write down the formula:

$$N = 5{,}781K + 1{,}005a - 2{,}001b + 997c - Q - \frac{K+R}{3}$$

Because a = 0, b = 25, c = 0, and 3Q + R = 0, so:

Q = 0 and R = 0

The data first gives us:

$$N = 5{,}781K + 0 - 50{,}025 + 0 + 0 - \frac{K+0}{3}$$

The data then shows that the correspondences of R = 0 are the least K = 9, $-\frac{K+R}{3} = -3$, and 5,781K = 52,029. Thus we have:

N = 52,029 − 50,025 − 3 = 2,001

Question 3

Find the least N so that:

- Upon division by 23, we get a remainder of 1
- Upon division by 26, we get a remainder of 1
- Upon division by 29 we also get a remainder of 1

Solution

We already know the formula:

$$N = 5{,}781K + 1{,}005a - 2{,}001b + 997c - Q - \frac{K+R}{3}$$

Because a = 1, b = 1, c = 1, and 3Q + R = 0, so:

Q=0 and R=0

The data first gives us:

$$N = 5{,}781K + 1{,}005 - 2{,}001 + 997 - 0 - \frac{K+0}{3}$$

$$= 5{,}781K + 1 - \frac{K+0}{3}$$

In the data, we have R = 0 corresponding to the least K_1 = 0 or K_2 = 3. We deduce:

$N_1 = 1$ or $N_2 = 17{,}343$

Question 4

Find the least N so that:

- Upon division by 23, we get a remainder of 22
- Upon division by 26, we get a remainder of 25
- Upon division by 29, we also get a remainder of 28

Solution

We first write down the formula:

$$N = 5,781K + 1,005a - 2,001b + 997c - Q - \frac{K+R}{3}$$

Because a = 22, b = 25, c = 28, and 3Q + R = 6, so:

Q=2 and R=0

The data first gives us:

$$N = 5,781K + 22,110 - 50,025 + 27,916 - 2 - \frac{K+0}{3}$$

$$= 5,781K - 1 - \frac{K+0}{3}$$

The data then gives us:

R = 0 corresponding to least K = +3, $-\frac{K+R}{3} = -1$, and 5,781K = 17,343

Ancient and Modern Mathematics

Therefore, we have:

$$N = 17{,}343 - 1 - 1 = 17{,}341$$

Question 5

Find the least N so that:

- Upon division by 23, we get a remainder of 0
- Upon division by 26, we get a remainder of 0
- Upon division by 29 we also get a remainder of 0

Solution

We first write down the formula:

$$N = 5{,}781K + 1{,}005a - 2{,}001b + 997c - Q - \frac{K+R}{3}$$

Because a = 0, b = 0, c = 0, and 3Q + R = 0, so:

$$Q = 0 \text{ and } R = 0$$

Therefore, the data gives us:

$$N = 5{,}781K + 0 - 0 + 0 - 0 - \frac{K+0}{3}$$

We then have R=0 corresponding to the least $K_1 = 0$ or $K_2 = 3$.

Therefore, we get:

$$N_1 = 0 \text{ or } N_2 = 17{,}342$$

Question 6

Find the least N so that:

- Upon division by 23, we get a remainder of 7
- Upon division by 26, we get a remainder of 13
- Upon division by 23, we get a remainder of 18

Solution

We first write down the formula:

$$N = 5{,}781K + 1{,}005a - 2{,}001b + 997c - Q - \frac{K+R}{3}$$

Because a = 7, b = 13, c = 18, and 3Q + R = 18 – 7, so:

Q = 3 and R = 2

The data first gives us:

$$N = 5{,}781K + 7{,}035 - 26{,}013 + 17{,}946 - 3 - \frac{K+2}{3}$$

$$= 5{,}781K + 24{,}981 - 26{,}016 - \frac{K+2}{3}$$

$$= 5{,}781K - 1{,}035 - \frac{K+2}{3}$$

We then have: R = 2 corresponding to K = 1, $-\frac{K+2}{3} = -1$, and 5,781K = 5,781. Thus, we get:

N = 5,781 – 1,036 = 4,745

Problem 10: The Improved Problem of Sunzi to Inspect Corps

Three allied nations named Nation A, Nation B, and Nation C join a military exercise at an unknown zone. By agreement, each nation sends the same amount of troops to that zone. This amount would be within a corps size. Prior to participating in the exercise maneuver, they must make their own formation and parade in front of the field commander assigned by the agreement. Nation A arranges a formation by using a platoon of thirty-one soldiers to make each row; the last row has only a ($0 \leq a \leq 30$) soldiers. Nation B does the same manner by using a platoon of thirty-four soldiers to form each row; the last row has only b ($0 \leq b \leq 33$) soldiers. Nation C also does the same thing by using a platoon of thirty-seven soldiers to form each row; the last row has only c soldiers ($0 \leq c \leq 36$). How many soldiers does each nation send to join that military exercise?

Solution

We first find the least common multiple of 31, 34, and 37 as: 31 x 34 x 37 = 38,998. We then set N as the amount of troops sent by each nation to the exercise zone. Then we have the total complete rows of the Nation A formation:

$$\frac{N-a}{31} = K_a$$

We have the total complete rows of the Nation B formation:

$$\frac{N-b}{34} = K_b$$

We have the total complete rows of the Nation C formation:

$$\frac{N-c}{37} = K_c$$

We recognize: $K_a \geq K_b \geq K_c$. We then establish the relationships as follows:

A. Set 38,998 as the common denominator of $\frac{N-a}{31} = K_a$, $\frac{N-b}{34} = K_b$, and $\frac{N-c}{37} = K_c$. We have:

$$\frac{1{,}258(N-a)}{38{,}998} = K_a, \quad \frac{1{,}147(N-b)}{38{,}998} = K_b, \quad \frac{1{,}054(N-c)}{38{,}998} = K_c$$

B. Like the previous problems, we must find out the least coefficient of N. So we consider 1,258N of K_a, 1,147N of K_b, and 1,054N of K_c. Then we make two relationships by subtracting:

1,258N − 1,147N = 111N (1)

and

1,147N − 1,054N = 93N (2)

If we multiply (1) by 5 and (2) by 6, we have:

5(1,258N − 1,147N) = 5 x 111N = 555N (3)

and

6(1,147N − 1,054N) = 6 x 93N = 558N (4)

If we subtract (4) by (3), we have:

11 x 1,147N − 5 x 1,258N − 6 x 1,054N = 3N (5)

Thus, we get 3 as the least coefficient of N. Based on (5), we can establish the formulas as follows:

$$\frac{11 \times 1,147(N-b) - 5 \times 1,258(N-a) - 6 \times 1,054(N-c)}{38,998} = 11K_b - 5K_a - 6K_c = K$$

Hence, we deduce:

$$N = \frac{38,998K + 12,617b - 6,290a - 6,324c}{3}$$

Or we can write:

$$N = \frac{38,997K + K + 12,618b - b - 6,291a + a - 6,324c}{3}$$

Finally, we have the formula between N and a, b, c, and K:

$$N = 12{,}999K + 4{,}206b - 2{,}097a - 2{,}108c + \frac{K+a-b}{3}$$

Set: $3Q + R = a - b$, and we know Q and R eventually

$-11 \leq Q \leq 10$ and $-2 \leq R \leq 2$

$$N = 12{,}999K + 4{,}206b - 2{,}097a - 2{,}108c + Q + \frac{K+R}{3}$$

$(-9 \leq K \leq 12)$

Remark

The solution is the same manner of the previous problems. The following is the data of problem 10 with the formula:

$$N = 12{,}999K + 4{,}206b - 2{,}097a - 2{,}108c + Q + \frac{K+R}{3}$$

Ancient and Modern Mathematics

b	4,206b	a	−2,097a	c	−2,108c	R	K	$\frac{K+R}{3}$	12,999K
1	4,206	1	−2,097	1	−2,108	−2	−7	−3	−90,993
2	8,412	2	−4,194	2	−4,216	−2	−4	−2	−51,996
3	12,618	3	−6,291	3	−6,324	−2	−1	−1	−12,999
4	16,824	4	−8,388	4	−8,432	−2	2	0	25,998
5	21,030	5	−10,485	5	−10,540	−2	5	1	64,995
6	25,236	6	−12,582	6	−12,648	−2	8	2	103,992
7	29,442	7	−14,679	7	−14,756	−2	11	3	142,989
8	33,648	8	−16,776	8	−16,864	−1	−8	−3	−103,992
9	37,854	9	−18,873	9	−18,972	−1	−5	−2	−64,995
10	42,060	10	−20,970	10	−21,080	−1	−2	−1	−25,998
11	46,266	11	−23,067	11	−23,188	−1	1	0	12,999
12	50,472	12	−25,164	12	−25,296	−1	4	1	51,996
13	54,678	13	−27,261	13	−27,404	−1	7	2	90,993
14	58,884	14	−29,358	14	−29,512	−1	10	3	129,990
15	63,090	15	−31,455	15	−31,620	0	−9	−3	−116,991
16	67,296	16	−33,552	16	−33,728	0	−6	−2	−77,994
17	71,502	17	−35,649	17	−35,836	0	−3	−1	−38,997
18	75,708	18	−37,746	18	−37,944	0	0	0	0
19	79,914	19	−39,843	19	−40,052	0	3	1	38,997
20	84,120	20	−41,940	20	−42,160	0	6	2	77,994
21	88,326	21	−44,037	21	−44,268	0	9	3	116,991
22	92,532	22	−46,134	22	−46,376	0	12	4	155,988
23	96,738	23	−48,231	23	−48,484	1	−7	−2	−90,993
24	100,944	24	−50,328	24	−50,592	1	−4	−1	−51,996
25	105,150	25	−52,425	25	−52,700	1	−1	0	−12,999
26	109,356	26	−54,522	26	−54,808	1	2	1	25,998
27	113,562	27	−56,619	27	−56,916	1	5	2	64,995

28	117,768
29	121,974
30	126,180
31	130,386
32	134,592
33	138,798

28	-58,716
29	-60-813
30	-62,910

28	-59,024
29	-61,132
30	-63,240
31	-65,348
32	-67,456
33	-69,564
34	-71,672
35	-73,780
36	-75,888

1	8	3	103,992
1	11	4	142,989
2	-8	-2	-103,992
2	-5	-1	-64,995
2	-2	0	-25,998
2	1	1	12,999
2	4	2	51,996
2	7	3	90,993
2	10	4	129,990

Exercises

Question I

Find the least N so that:

- Upon division by 31, we get a remainder of 0
- Upon division by 34, we get a remainder of 33
- Upon division by 37, we get a remainder of 0

Solution

We first write down the formula:

$$N = 12,999K + 4,206b - 2,097a - 2,108c + Q + \frac{K+R}{3}$$

Because a = 0, b = 33, c = 0, and 3Q + R = a − b = −33, so:

Q = −11 and R = 0

The data first gives us:

$$N = 12{,}999K + 138{,}798 - 0 - 0 - 11 + \frac{K+0}{3}$$

We then have:

$R = 0$ corresponding to the least $K = -9$, $\frac{K+R}{3} = -3$, and $12{,}999K = -116{,}991$.

Therefore, we get:

$$N = -116{,}991 + 138{,}798 - 11 - 3 = 21{,}793$$

Question 2

Find the least N so that:

- Upon division by 31, we get a remainder of 30
- Upon division by 34, we get a remainder of 0
- Upon division by 37, we get a remainder of 36

Solution

We first have to write down the formula:

$$N = 12{,}999K + 4{,}206b - 2{,}097a - 2{,}108c + Q + \frac{K+R}{3}$$

Because $a = 30$, $b = 0$, $c = 36$, and $3Q + R = a - b = 30 - 0 = 30$, so:

$Q = 10$ and $R = 0$

The data first gives us:

$$N = 12{,}999K + 0 - 62{,}910 - 75{,}888 + 10 + \frac{K+0}{3}$$
$$= 12{,}999K - 138{,}788 + \frac{K+0}{3}$$

We then recognize:

R = 0 corresponding to the least K=12, $\frac{K+R}{3} = 4$, and 12,999K = 155,988.

Therefore, we get the least N as:

N = –155,998 – 138,798 + 4 = 17,104

Question 3

Find the least N so that:

- Upon division by 31, we get a remainder of 0
- Upon division by 34, we get a remainder of 0
- Upon division by 37, we get a remainder of 0

Solution

Consider the formula:

$$N = 12{,}999K + 4{,}206b - 2{,}097a - 2{,}108c + Q + \frac{K+R}{3}$$

Because a = 0, b = 0, c = 0, and 3Q + R = 0, so:

Q = 0 and R = 0

Therefore, we have:

$$N = 12,999K + 0 - 0 - 0 + 0 + \frac{K+0}{3}$$

In the data, we have:

R = 0 corresponding to K=0 or K=3.

Therefore, we have:

N = 0 or N = 38,998

Question 4

Find the least N so that:
- Upon division by 31, we get a remainder of 1
- Upon division by 34, we get a remainder of 1
- Upon division by 37, we get a remainder of 1

Solution

We already recognize the formula:

$$N = 12,999K + 4,206b - 2,097a - 2,108c + Q + \frac{K+R}{3}$$

Because a = 1, b = 1, c = 1, and 3Q + R = 1 − 1 = 0, so:

Q = 0 and R = 0

Therefore, we get:

$$N = 12{,}999K + 4{,}206 - 2{,}097 - 2{,}108 + 0 + \frac{K+0}{3}$$
$$= 12{,}999K + 1 + \frac{K+0}{3}$$

The data gives us:

R = 0 corresponding to $K_1 = 0$ or $K_2 = 3$, $\frac{K_1 + 0}{3} = 0$,

or $\frac{K_2 + 0}{3} = 1$

Therefore, we have:

N = 1 or N = 38,999

Question 5

Find the least N so that:

- Upon division by 31, we get a remainder of 30
- Upon division by 34, we get a remainder of 33
- Upon division by 37, we get a remainder of 36

Solution

We first write down the formula:

$$N = 12,999K + 4,206b - 2,097a - 2,108c + Q + \frac{K+R}{3}$$

According to the hypothesis, we have:

a = 30, b = 33, c = 36, and 3Q + R = 30 − 33 = −3, so:

Q = −1 and R = 0

The data first gives us:

$$N = 12,999K + 138,798 - 62,910 - 75,888 - 1 + \frac{K+0}{3}$$
$$= 12,999K - 1 + \frac{K+0}{3}$$

We then get:

R = 0 corresponding to K = 3, $\frac{K+R}{3} = 1$, and 12,999K = 38,997

Therefore, we get the least N as below:

N = 38,997 − 1 + 1 = 38,997.

Question 6

Find the least N so that:
- Upon division by 31, we get a remainder of 7
- Upon division by 34, we get a remainder of 24
- Upon division by 37, we get a remainder of 3

Solution

Consider the formula:

$$N = 12{,}999K + 4{,}206b - 2{,}097a - 2{,}108c + Q + \frac{K+R}{3}$$

According to the hypothesis, we have:

a = 7, b = 24, c = 3, and 3Q + R = 7 − 24 = −17, so:

Q = −5 and R = −2

The data first gives us:

$$N = 12{,}999K + 100{,}944 - 14{,}679 - 6{,}324 - 5 + \frac{K-2}{3}$$
$$= 12{,}999K + 79{,}936 + \frac{K-2}{3}$$

The data then gives us:

R = −2 corresponding to the least K = −4, $\frac{K+R}{3} = -2$, and 12,999K = −51,996.

Therefore, we deduce:

N = −51,996 + 79,936 − 2 = 27,938

Problem 11: The Improved Problem of Sunzi to Inspect a Group of People

A group of tourists travels to Europe. Following their schedule, we recognize the following. The first day at Vienna, the whole group joins a concert at the national theater where the main section was reserved for this group. This section has many rows of fifteen seats, and the group occupies almost the rows, except the last row, which consists of a ($0 \leq a < 15$) occupied seats. The second day in Paris, the group checks in at the multiple-floor hotel. Each floor consists of twenty-five rooms. They occupy almost the whole hotel, and no rooms are left vacant, except the last floor at the top. There are only b ($0 \leq b < 25$) occupied rooms. On the third day, the group takes the high-speed train to go to Rome. At the train station, they are due to share rooms with the other groups of tourists. The conductors allow thirty-five people to board each coach. The remainder c ($0 \leq c < 35$) people have to board the last coach. How many people are in this group of tourists?

Solution: Part I

We first find the least common multiple of 15, 25, and 35 as: $15 \times 25 \times 35 = 13{,}125 = 5^3 \times 105 \rightarrow 5 \times 105 = 525$. We set N as the number of people in the group. Then we have the total complete rows of seats occupied by that group:

$$\frac{N-a}{15} = K_a$$

We have the total complete floors of 25 rooms occupied by that group:

$$\frac{N-b}{25} = K_b$$

We have the total complete coaches of 35 people of that group:

$$\frac{N-c}{35} = K_c$$

We recognize: $K_a \geq K_b \geq K_c$. Now we establish the formula of N as follows. We set 525 as the common denominator of $\frac{N-a}{15} = K_a$, $\frac{N-b}{25} = K_b$, and $\frac{N-c}{35} = K_c$. We have:

$$\frac{35(N-a)}{525} = K_a, \quad \frac{21(N-b)}{525} = K_b, \quad \frac{15(N-c)}{525} = K_c$$

In considering these relationships K_a, K_b, and K_c, we can obtain the formula of N as:

$$\frac{21(N-b) + 15(N-c) - 35(N-a)}{525} = K_b + K_c - K_a = K$$

Hence, we deduce:

N − 21b − 15c + 35a = 525K

Finally, we obtain the formula of the least N: N = 525K + 15c + 21b − 35a (−1 ≤ K ≤ 1).

Solution: Part 2

Consider the following relationships we get:

$$*\frac{N-a}{15} = K_a \rightarrow \frac{N-a}{5} = 3K_a \rightarrow$$
$$*\frac{N-b}{25} = K_b \rightarrow \frac{N-b}{5} = 5K_b \rightarrow \begin{bmatrix} b-a = 5(3K_a - 5K_b) = 5K_1 \, (-2 \leq K_1 \leq 4) \\ c-b = 5(5K_b - 7K_c) = 5K_2 \, (-4 \leq K_2 \leq 6) \\ c-a = 5(3K_a - 7K_c) = 5K_3 \, (-2 \leq K_3 \leq 6) \end{bmatrix}$$
$$*\frac{N-c}{35} = K_c \rightarrow \frac{N-c}{5} = 7K_c \rightarrow$$

Finally, we have three important conditions of possibility:

- |b − a| = 5|K_1| (−2 ≤ K_1 ≤ 4) or |b − a| must be divisible by 5.
- |c − b| = 5|K_2| (−4 ≤ K_2 ≤ 6) or |c − b| must be divisible by 5.
- |c − a| = 5|K_3| (−2 ≤ K_3 ≤ 6) or |c − a| must be divisible by 5.

The problem could be solved only if a, b, or c would satisfy those conditions above.

Solution: Part 3

The data for the problem 11 with the formula:

c	15c
1	15
2	30
3	45
4	60
5	75
6	90
7	105
8	120
9	135
10	150
11	165
12	180
13	195
14	210
15	225
16	240
17	255
18	270
19	285
20	300
21	315
22	330
23	345
24	360
25	375
26	390

b	21b
1	21
2	42
3	63
4	84
5	105
6	126
7	147
8	168
9	189
10	210
11	231
12	252
13	273
14	294
15	315
16	336
17	357
18	378
19	399
20	420
21	441
22	462
23	483
24	504

a	−35a
1	−35
2	−70
3	−105
4	−140
5	−175
6	−210
7	−245
8	−280
9	−315
10	−350
11	−385
12	−420
13	−455
14	−490

$N = 525K + 15c + 21b - 35a$

$-1 \leq K \leq 1$

$0 \leq a \leq 14$

$0 \leq b \leq 24$

$0 \leq c \leq 34$

The least $N : 0 \leq N \leq 525$

27	405
28	420
29	435
30	450
31	465
32	480
33	495
34	510

Exercises

Question I

Find the least N so that:

- Upon division by 15, we get a remainder of 0
- Upon division by 25, we get a remainder of 0
- Upon division by 35, we get a remainder of 0

Solution

We first write down the formula:

$$N = 525K + 15c + 21b - 35a$$

Because $a = 0$, $b = 0$, and $c = 0$, we have:

$$N = 525K$$

Therefore, the least N corresponds to the least K = 0 or K = 1. Finally, we have:

$$N = 0 \text{ or } N = 525$$

Note: To know the least N = 525, which is true or not, we have to check the conditions of a, b, and c:

- $|b - a| = 5.|K_1|$ $|c - b| = 5.|K_2|$ $|c - a| = 5.|K_3|$
- $|0 - 0| = 5.|0|$ $|0 - 0| = 5.|0|$ $|0 - 0| = 5.|0|$

These three conditions are satisfied, so N = 525 is a true answer.

Question 2

Find the least N so that:

- Upon division by 15, we get a remainder of 1
- Upon division by 25, we get a remainder of 1
- Upon division by 35, we get a remainder of 1

Solution

Prior to find the least N, we must check the conditions:

- $|b - a| = 5.|K_1|$ $|c - b| = 5.|K_2|$ $|c - a| = 5.|K_3|$
- $|1 - 1| = 5.|0|$ $|1 - 1| = 5.|0|$ $|1 - 1| = 5.|0|$
- $|0| = 5.|0|$ $|0| = 5.|0|$ $|0| = 5.|0|$

These three conditions are satisfied, so the problem can be solved. Now we write down the formula:

$$N = 525K + 15c + 21b - 35a$$

Because $a = 1$, $b = 1$, and $c = 1$, we have:

$$N = 525K + 15 + 21 - 35 = 525K + 1$$

Therefore, the least N corresponds to $K = 0$ or $K = 1$, and we have the answer:

$$N = 1 \text{ or } N = 526$$

Question 3

Find the least N so that:
- Upon division by 15, we get a remainder of 14
- Upon division by 25, we get a remainder of 24
- Upon division by 35, we get a remainder of 34

Solution

We first check the conditions of a, b, and c:
- $|b - a| = 5.|K_1|$, $|c - b| = 5.|K_2|$, $|e - a| = 5.|K_3|$
- $|24 - 14| = |10|$, $|34 - 24| = |10|$, $|34 - 14| = |20|$
- $|10| = 5.|2|$, $|10| = 5|2|$, $|20| = 5.|4|$

These conditions are satisfied, so we can solve the problem:

$$N = 525K + 15c + 21b - 35a$$

Because $a = 14$, $b = 24$, and $c = 34$, so we have:

$$N = 525K + 510 + 504 - 490 = 525K + 524$$

The least N corresponds to the least K=0, so we have: N = 524

Question 4

Find the least N so that:

- Upon division by 15, we get a remainder of 0
- Upon division by 25, we get a remainder of 20
- Upon division by 35, we get a remainder of 30

Solution

We first check the conditions of a, b, and c:

- $|b - a| = 5.|K_1|$, $|c - b| = 5.|K_2|$, $|c - a| = 5.|K_3|$
- $|20 - 0| = |20|$, $|30 - 20| = |10|$, $|30 - 0| = |30|$
- $|20| = 5.|4|$, $|10| = 5|2|$, $|30| = 5.|6|$

These conditions are satisfied, so we can solve the problem:

$$N = 525K + 15c + 21b - 35a$$

Because a = 0, b = 20, and c = 30, we have:

$$N = 525K + 450 + 420 - 0$$

Or we have:

$$N = 525 K + 870$$

The least N here corresponds to the least K = −1.

Thus, we deduce:

$$N = -525 + 870 = 345$$

Question 5

Find the least N so that:

- Upon division by 15, we get a remainder of 10
- Upon division by 25, we get a remainder of 0
- Upon division by 35 we get a remainder of 0

Solution

We first check the conditions of a, b, and c:

- $|b - a| = 5.|K_1|$, $|c - b| = 5.|K_2|$, $|c - a| = 5.|K_3|$
- $|0 - 10| = |-10|$, $|0 - 0| = |0|$, $|0 - 10| = |-10|$
- $|-10| = 5.|-2|$, $|0| = 5.|0|$, $|-10| = 5.|-2|$

These conditions are satisfied, so we can solve the problem:

$$N = 525K + 0 + 0 - 350$$

The least N here corresponds to the least $K = 1$, so we have:

$$N = 525K - 350 = 175$$

Question 6

Find the least N so that:

- Upon division by 15, we get a remainder of 12
- Upon division by 25, we get a remainder of 17
- Upon division by 35, we get a remainder of 32

Solution

We first check the conditions of a, b, and c:

- $|b - a| = 5.|K_1|, |c - b| = 5.|K_2|, |c - a| = 5.|K_3|$
- $|17 - 12| = 5, |32 - 17| = |15|, |32 - 12| = |20|$
- $|5| = 5.|1|, |15| = 5.|3|, |20| = 5.|4|$

These conditions are satisfied, so the problem can be solved:

$$N = 525K + 15c + 21b - 35a = 525K + 480 + 357 - 420$$
$$= 525K + 417$$

The least N corresponds to the least $K = 0$, so we get:

$$N = 417$$

Question 7

Find the least N so that:

- Upon division by 15, we get a remainder of 8
- Upon division by 25, we get a remainder of 15
- Upon division by 35, we get a remainder of 4

Solution

We first check the conditions of a, b, and c:

- $|b - a| = 5.|K_1|, |c - b| = 5.|K_2| |c - a| = 5.|K_3|$
- $|15 - 8| = |7|, |4 - 15| = |-11| |4 - 8| = |1 - 4|$

We recognize that $|7|$, $|-11|$, and $|-4|$ are not the multiples of 5, so the conditions of possibility are not satisfied. Therefore, the problem can't be solved. If we ignore these and try to find the least N to see, what happens? We already know the formula:

$$N = 525K + 15c + 21b - 35a$$

Because a = 8, b = 15, and c = 4, we get:

$$N = 525K + 60 + 315 - 280 = 525K + 95$$

The least N corresponds to the least K = 0, so we get:

$$N = 95$$

To verify the result, we let 95 be divided by 15. We get a remainder of 5, not 8 as the hypothesis said. Then we let 95 be divided by 25. We get a remainder of 20, not 15 as the hypothesis said. Finally, we let 95 be divided by 35. We get a remainder of 25, not 4 as the hypothesis said. So N = 95 is not a true answer for a = 8, b = 15, and c = 4. That means the problem has no answer.

Question 8

Find the least N so that:

- Upon division by 15, we get a remainder of 5
- Upon division by 25, we get a remainder of 20
- Upon division by 35, we get a remainder of 25

Solution

We first check the conditions of possibility a, b, and c as:

- $|b - a| = 5.|K_1|$, $|c - b| = 5.|K_2|$, $|c - a| = 5.|K_3|$
- $|20 - 5| = |15|$, $|25 - 20| = |5|$, $|25 - 5| = |20|$
- $|15| = 5.|3|$ $|5| = 5.|1|$ $|20| = 5.|4|$

These conditions are satisfied, so the problems can be solved as follows:

$$N = 525K + 15c + 21b - 35a$$

Because a = 5, b = 20, and c = 25, the data gives us:

$$N = 525K + 375 + 420 - 175 = 525K + 620$$

The least N here corresponds to K = −1, so we get:

$$N = -525 + 650 = 95$$

Chapter 4

Problem 12: The Triangle with Two Equal Bisectors

Consider a triangle ABC in which two bisectors of angle B and C at the base BC are equal to one another.

1. Prove that ABC is an isosceles triangle.
2. If the isosceles triangle ABC has the base BC = a, what are the possibility conditions of those equal bisectors for ABC to exist?

Demonstration

Question I

Suppose the triangle ABC in which two bisectors BM and CN of angle B and angle C, respectively, have equal lengths, as shown by the figure below. In ABC, the point 0 is the intersection of these two bisectors, BM and CN. The points M and N are the intersections of BM and CN with AC and AB, respectively. Let OA be joined. Then OA is also a bisector of angle A of ABC; OA produced cuts BC at H.

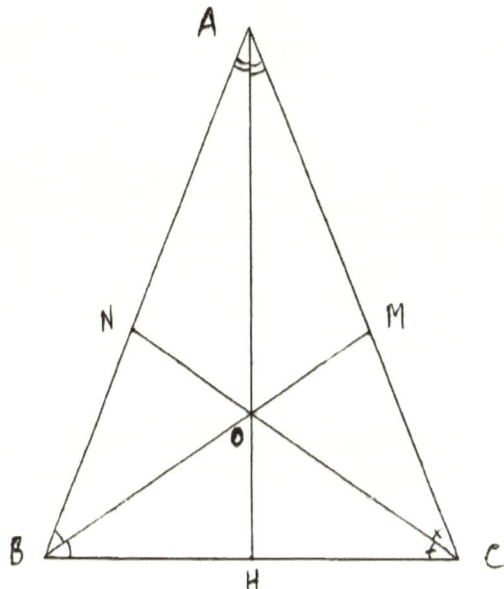

Now everything is set. We first consider triangle ABM with the base BM and OA as a bisector of angle A. Then we consider triangle CBM with the base BM and OC as a bisector of angle C. According to the proposition 3 in Book VI of Euclid, we have:

$$\frac{OM}{AM} = \frac{OB}{AB} = \frac{OM+OB}{AM+AB} = \frac{BM}{AM+AB} \quad (1)$$

and

$$\frac{OM}{CM} = \frac{OB}{BC} = \frac{OM+OB}{CM+BC} = \frac{BM}{CM+BC} \quad (2)$$

Again, we first consider triangle ACN with the base CN and OA as a bisector of angle A. Then we consider triangle BCN with the base CN and OB as a bisector of angle B. We also have:

$$\frac{ON}{AN} = \frac{OC}{AC} = \frac{ON+OC}{AN+AC} = \frac{CN}{AN+AC} \quad (3)$$

and

$$\frac{ON}{BN} = \frac{OC}{BC} = \frac{ON+OC}{BN+BC} = \frac{CN}{BN+BC} \quad (4)$$

From (1) and (2) we deduce:

$$BM = \frac{OM.(AM+AB)}{AM} \quad (a)$$

$$BM = \frac{OM.(CM+BC)}{CM} \quad (b)$$

$$BM = \frac{OB.(AM+AB)}{AB} \quad (c)$$

and

$$BM = \frac{OB.(CM+BC)}{BC} \quad (d)$$

From (3) and (4), we deduce:

$$CN = \frac{ON.(AN+AC)}{AN} \quad (a')$$

$$CN = \frac{ON.(BN+BC)}{BN}, \quad (b')$$

$$CN = \frac{OC.(AN+AC)}{AC} \quad (c')$$

and

$$CN = \frac{OC.(BN+BC)}{BC} \quad (d')$$

From (a) and (b), we can write:

$$BM = \frac{OM.(AM+AB)}{AM} = \frac{OM.(CM+BC)}{CM}$$
$$= \frac{OM.(AM+AB)+OM.(CM+BC)}{AM+CM}$$
$$= \frac{OM.(AB+AM+CM+BC)}{AM+CM}$$

But AM + CM = AC, we get:

$$BM = \frac{OM.(AB+AC+BC)}{AC} \quad (\alpha)$$

From (c) and (d), we can write:

$$BM = \frac{OB \cdot (AM + AB)}{AB} = \frac{OB \cdot (CM + BC)}{BC}$$

$$= \frac{OB \cdot (AM + AB) + OB \cdot (CM + BC)}{AB + BC}$$

$$= \frac{OB \cdot (AB + AM + CM + BC)}{AB + BC}$$

But: AM + CM = AC, we get:

$$BM = \frac{OB(AB + AC + BC)}{AB + BC} \quad (\beta)$$

From (a') and (b'), we can write:

$$CN = \frac{ON \cdot (AN + AC)}{AN} = \frac{ON \cdot (BN + BC)}{BN}$$

$$= \frac{ON \cdot (AN + AC) + ON \cdot (BN + BC)}{AN + BN}$$

$$= \frac{ON \cdot (AN + BN + AC + BC)}{AN + BN}$$

But:

AN + BN = AB, we get:

$$CN = \frac{ON \cdot (AB + AC + BC)}{AB} \quad (\alpha')$$

From (c′) and (d′), we can write:

$$CN = \frac{OC.(AN+AC)}{AC} = \frac{OC.(BN+BC)}{BC}$$
$$= \frac{OC.(AN+AC)+OC.(BN+BC)}{AC+BC}$$
$$= \frac{OC.(AN+BN+AC+BC)}{AC+BC}$$

But: AN + BN = AB, we get:

$$CN = \frac{OC.(AB+AC+BC)}{AC+BC} \quad (\beta')$$

Because the hypothesis gives us BM = CN, we deduce from (α) and (α′):

$$\frac{OM.(AB+AC+BC)}{AC} = \frac{ON.(AB+AC+BC)}{AB}$$

If we simplify the relationship above by eliminating the common factor (AB + AC + BC), we get:

$$\frac{OM}{AC} = \frac{ON}{AB} = R_1 \quad (I)$$

Also from (β) and (β′), we deduce:

$$\frac{OB.(AB+AC+BC)}{AB+BC} = \frac{OC.(AB+AC+BC)}{AC+BC}$$

If we simplify the relationship above by eliminating the common factor (AB + AC + BC), we get:

$$\frac{OB}{AB+BC} = \frac{OC}{AC+BC} = R_2 \quad \text{(II)}.$$ From (I), we deduce:

$$OM = R_1 \cdot AC \text{ and } ON = R_1 \cdot AB$$

From (II), we deduce:

$$OB = R_2 \cdot (AB + BC) \text{ and } OC = R_2 \cdot (AC + BC)$$

But:

$$OM + OB = BM, ON + OC = CN, \text{ and } BM = CN$$

Thus, we have:

$$R_1 \cdot AC + R_2 \cdot (AB + BC) = R_1 \cdot AB + R_2 \cdot (AC + BC)$$

We deduce:

$$R_1 \cdot AC + R_2 \cdot (AB + BC) - R_1 \cdot AB - R_2 \cdot (AC + BC) = 0$$

Or we can rearrange and write:

$$R_1 \cdot (AC - AB) + R_2 \cdot (AB + \cancel{BC} - AC - \cancel{BC}) = 0$$

Or we can rewrite:

$$R_1 \cdot (AC - AB) + R_2 \cdot (AB - AC) = 0$$

We can rearrange and get:

$$R_2 \cdot (AB - AC) - R_1 \cdot (AB - AC) = 0$$

Finally, we have:

$$(R_2 - R_1)(AB - AC) = 0$$

But $R_2 \neq R_1$. Thus, we get:

$$AB - AC = 0$$

Therefore:

$$AB = AC$$

This proves the triangle ABC is an isosceles triangle.

Question 2

If the isosceles triangle ABC has the base BC = a and the equal sides AB and AC are variable, so are the angles B and C. The equal angles B and C vary from 0 to 90 degrees. Because ABC is an isosceles triangle, A and O always locate at the bisector AH of BC. (The perpendicular bisects BC at H.) Therefore, when angle B and C become 0 degrees, the triangle ABC becomes a flat triangle. A and O then locate at H. In this case:

$$AB = AC = BO = OC = \frac{BC}{2} = \frac{a}{2}$$

Thus, we deduce:

$$BM = \frac{OB.(AB+AC+BC)}{AB+BC}$$
$$= \frac{\frac{a}{2}.\left(\frac{a}{2}+\frac{a}{2}+a\right)}{\frac{a}{2}+a} = \frac{a^2}{\frac{3a}{2}}$$

We have:
$$BM = CN = \frac{2a}{3}$$

When angles B and C increase and reach 90 degrees, A locates at infinite. In this case, the bisectors BM and CN make the angles MBC and NCB with BC, which equal to 45 degrees. Therefore, the triangles CBM and BCN are isosceles and right-angled triangles. The hypotenuses BM and CN of triangles CBM and BCN are therefore:

$$\overline{BM}^2 = \overline{CN}^2 = \overline{BC}^2 + \overline{CM}^2 = \overline{BC}^2 + \overline{BN}^2 \quad \text{(Book 1, Pr 47)}$$

But:
$$BC = CM = BN = a$$

Therefore, we get:
$$\overline{BM}^2 = \overline{CN}^2 = a^2 + a^2 = 2a^2$$

We deduce:
$$BM = CN = a\sqrt{2}$$

Therefore, the possibility conditions of BM and CN for the isosceles triangle ABC to exist are below:

$$\frac{2a}{3} < l < a\sqrt{2}$$ (where l is the length of BM and CN).

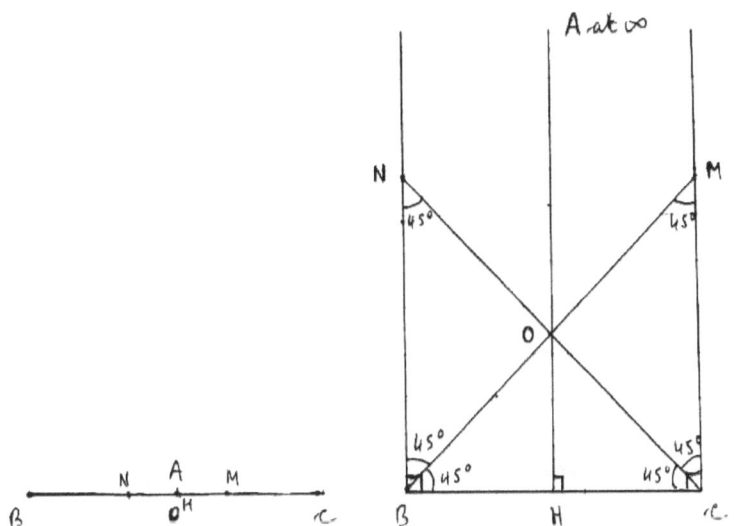

Problem 13: A Straight Line and a Circle Intersect at Two Points

Prove that a straight line has one point inside and one point outside a circle. It has two points common with the circle.

Demonstration

According to Euclid, a given straight line has one point inside and one point outside a circle. It has two points common with this circle. Euclid tacitly assumes it as an obvious truth. However, many geometers did not agree to it. Those refuters consist of two main groups:

- A group of geometers who refute it and interpolate that a given straight line and a given circle have more than two common points if they satisfy a certain criteria that we mentioned above
- Another group who consider Euclid's assumption be not obvious and try to prove it

To clarify this subject matter, the demonstration must go through three cases: refutation of first group, demonstration of second group, and demonstration of the author.

Question I: Refutation of First Group

The first group of geometers asserts that a given straight line will intersect a given circle at more than two points. They criticize it with disregard of definition and construction of the straight line and that of the circle. To show a straight line and a circle having more than two common points, they describe a so-called straight line and a so-called circle with a deformation, like a monster shape, because they cannot force the truth to respect their interpolation, but the reverse.

More importantly, their refutation is baseless and has no proof, so it is invalid. If we denounce someone of a wrongdoing with no evidence or proof, this defendant has the right to deny it with no evidence or proof either. Therefore, the tacit assumption of Euclid is still truthful. Yes, throughout twenty-three centuries, people worldwide have applied this truth in many trillions of times, but they haven't found any single case that shows a straight line and a circle having more than two common points. Thus, this refutation must be set aside.

Question 2: Demonstration of Second Group

Consider the circle (c) with center o and radius R and a straight line (r) with point A inside and one point B outside the circle. By definition of the circle, we have:

$$OA < R, OB > R$$

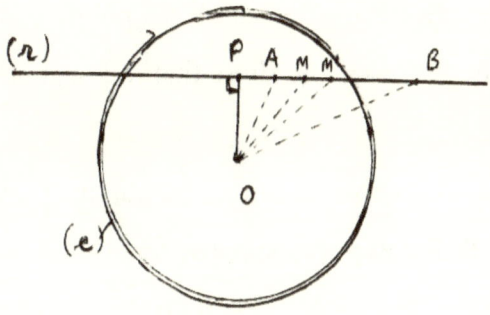

Draw OP perpendicular to the straight line (r). Then OP < OA, so OP is always less than R, and P is, therefore, within the circle (c). Now let us fix our attention on the finite segment AB of the straight line r. It can be divided into two parts:

- Containing all the points H for which OH < R (i.e., points inside c)
- Containing all the points K for which OK ≥ R (points outside c or on the circumference of c)

Thus remembering that, of two obliques from a given point to a given straight line, is greater the projection of which is greater, we can assert that all the points of the segment PB that precede a point inside c are inside c, and those that follow a point on the circumference of c or outside c are outside c.

Hence, by the postulate of Dedekind, there exists a point M on the segment PB such that all the points that precede it belong to the first part and those that follow it belong to the second part.

I say that M is common to the straight line r and the circle c, or OM = R. For example, suppose that OM < R. There will exist a segment (or length) σ less than the difference between R and OM. Consider the point M', one of those that follow M such that MM' is equal to σ. Then, because any side of a triangle is less than the sum of the other two, OM' < OM + MM'. But OM + MM' = OM + σ < R, OM' < R (which is absurd).

A similar absurdity would follow if we suppose that OM > R. Therefore, OM must be equal to R. It is immediately obvious that, corresponding to the point M on the segment PB that is common to r and c, another point on r has the same property, namely that which is symmetrical to M with respect to P. The proposition is proved.

Comments

- α) Follow up this demonstration. At the beginning, we find out a confusion as: "Draw OP perpendicular to the straight line r." How do we draw OP perpendicular to straight line r? Where does this construction come from? Of course, it comes from Proposition 12 in Book 1. Prior to drawing OP, we already accept that the circle (c) with center O cuts the straight line r at two points GE, so they use the less obvious truth, "A circle cuts a straight line utmost two points," to prove the same thing, "A circle cuts a straight line utmost two points." After that, they use Proposition 1.10 of Euclid to bisect the straight segment GE at P. But Proposition 1.10

cannot exist without Proposition 1.1, and Proposition 1.1 assumes that two circles cut utmost two points. This truth is much less obvious than the truth, "A straight line and a circle have two points common."

- β) Near the end of the demonstration, we find another confusion. Because any side of a triangle is less than the sum of the other two:

$$OM' < OM + MM'$$

Where does this proof comes from? It comes from Proposition 1.20. This proposition cannot exist without the premised propositions that strictly relate to it:

- Proposition 1.20 deduced from Proposition 1.19 and 1.5.
- Proposition 1.19 deduced from Proposition 1.18 and 1.5.
- Proposition 1.18 deduced from Proposition 1.16 and 1.3.
- Proposition 1.16 deduced from Proposition 1.10 and 1.3.
- Proposition 1.10 deduced from Proposition 1.9 and 1.1.
- Proposition 1.9 deduced from Proposition 1.8 and 1.3.
- Proposition 1.8 deduced from Proposition 1.7.
- Proposition 1.7 deduced from Proposition 1.5.
- Proposition 1.5 deduced from Proposition 1.3.

Concerning Proposition 1.3, we must distinguish it in two cases:

- **First Case: A Given Straight Line BC the less and a Cutoff Straight Line AD in AX** the greater have different extremities. (B and A are the extremities of BC and AD, respectively.) In this case, Proposition 1.3 relates to Proposition 1.2 and 1.1, which requires the construction of an equilateral triangle with the sides equal to AB. This construction, of course, is based on the assumption of two circles having two common points.
- **Second Case: A Given Straight Line AF the Less and a Cutoff Straight Line AG in AX** the greater have the same extremity A. Euclid forgot to mention this particular case. In this case, Proposition 1.3 has no equilateral triangle to construct because the given straight line AF (the less) and the cutoff straight line AG in AX (the greater) have the same extremity A. So Proposition 1.3 in this case is independent of Proposition 1.2 and 1.1.

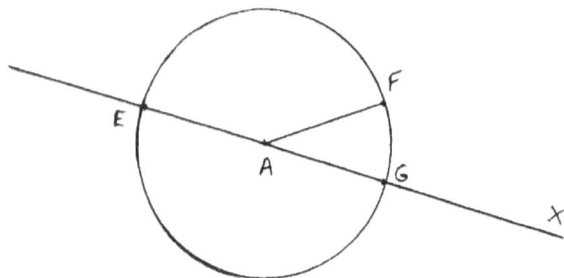

To get a cutoff straight line AG in AX, the greater, equal to AF, we only describe a circle at a point A as its center with the radius AF. Then we must accept "the circle cutting the straight line AX at two points G and E" as a truth.

Now we return to the demonstration of the commentators. Because the demonstration based on Proposition 1.20, which relates strictly to a chain of the premised propositions, "Proposition 1.19, 1.18 … 1.10 … 1.3 (second case), and 1.1." For the second case of Proposition 1.3 to exist, the circle must cut the straight line at two points utmost. And for Proposition 1.1 to exist, two circles must meet in two points utmost.

Conclusion

Obviously, the commentators use the tacit assumption of Euclid to prove that of Euclid. This is a confusion, so they prove nothing.

Question 3: Demonstration of the Author

First, to demonstrate this tacit assumption, we must avoid using the premised propositions that relate to "a given straight line and a given circle having two common points." Second, we consider a given circle (c) with center O and a radius R. With this circle, we must recognize four properties:

1. According to Postulate 3, we can only draw one circle and one only from a center O with a radius R.
2. The circumference of a circle is a perfect curve with equal curvatures all around it, so the curvature has no sharp turn at any portions of the circle. This property shows the fallacy of the first group of geometers.
3. The circumference of a circle is also the locus of the points from which the distances to center O are equal to R.
4. The circumference of a circle divides its plane surface into two areas: "the enclosed area and the outside area of the circle."

According to the definition of a circle:

- Any points at the enclosed area have the distances from them to the center O being lesser than the radius R: OA < R (A, a point at the enclosed area).
- Any points at the outside area have the distances from them to the center O being greater than the radius R: OB > R (B, a point at the outside area).

Now we return to the demonstration. Considering a given circle (c) at center O with the radius R and a given straight line (S) having one point M inside and one point N outside of a circle (c), I say that the straight line MN produced and the circle (c) have utmost two common points.

Demonstration

First of all, we draw the figure (1) as below:

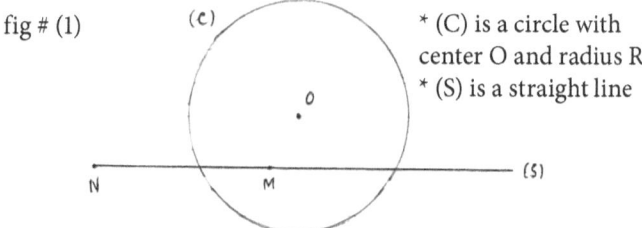

fig # (1)

* (C) is a circle with center O and radius R
* (S) is a straight line

Second, copy the figure (1) and rotate this copy 180° about straight line (S) then we get the figure (2):

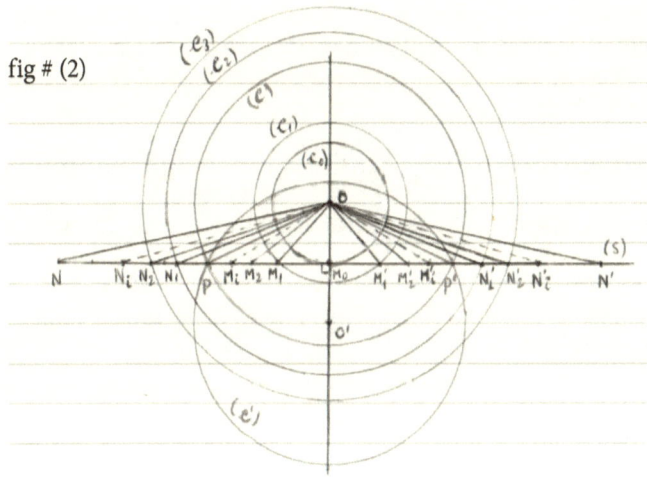

fig # (2)

The result of this rotation shows that the circle (C) and circle (C') are symmetrical one another with respect to the straight line (S) as a symmetric axis, therefore if we join 00', the straight line 00' is perpendicular to the straight line (S) at a point Mo.

1. Describe the circle (C_0) with center 0 and the radius OM_0, this circle touches the straight line (S) at M_0 only. So the entire straight line (S) locates outside the circle (C_0), except a touching point M_0.

If M locates at a point M_1 on the straight line (S), that means M_1 being outside the circle (C_0). Thus we have:

$OM_1 > OM_0$ (see the fourth property of a circle).

Describe the circle (C_1) with the center 0 and the radius OM_1, this circle takes OO' as it symmetrical axis. If we rotate the circle (C_1) to 180° about this axis 00', we get the correspondent point M_1' of M_1.

Because $OM_0 < OM_1 = OM_1'$ so the straight portion M_1M_1' of the straight line (S) is inside the circle (C_1), thus the other two portions M_1N and $M_1'N'$ are outside the circle (C_1)

If M moves to M_2 on the straight line (S) in the direction M_1N, that means M_2 being outside the circle (C_1) thus we have: $OM_2 > OM_1$ (fourth property of a circle).

All the results above show that:

$OM_2 > OM_1 > OM_0$

But the points M_0, M_1 and M_2 are inside circle (C), therefore we have:

$R > OM_2 > OM_1 > OM_0$ [R is the radius of (C)]

If M moves to any point Mi between point M_2 and point P inside the circle (C), with the same demonstration above we have: $R > OM_i > .... > OM_2 > OM_1 > OM_0$

2. Now, if a point M continues to move to the point N_1 outside the circle (C) we have:

$ON_1 > R$ (the fourth property of a circle).

Describe the circle (C_2) with center 0 and radius ON_1. If M continues to reach the point N_2 outside the circle (C_2) we have:

$ON_2 > ON_1 > R$.

If M continues to reach Ni then reach N, with the same demonstration we have:

$ON > ONi > ... > ON_2 > ON_1 > R$.

Summarize all the results above we obtain:

$ON > ONi > ... > ON_2 > ON_1 > R > OMi > ... > OM_2 > OM_1 > OM_o$.

3. We recognize that the point M when moves continuously from M_o inside circle (C) to N outside (C), in connection with center 0 it produces an infinite radii of circles (Ci) that gradually increase from the minimum OM_o to the maximum ON:

$ON > ONi > ... > ON_2 > ON_1 > R > OMi > ... > OM_2 > OM_1 > OM_o$.

These series of radii are divided into two parts by radius R of circle (C) and they correspond univocally to the points on the straight line (S) as below:

N Ni ...... N_2 N_1 P Mi .... M_2 M_1 M_o

All radii on the left side of R belong to the first part.

All radii on the right side of R belong to the second part.

Since every radii of the first part precede R and every radii of the second part follow R, therefore by the postulate of DedeKind there must exist a radius OP such that OP = R.

But P is on the straight segment M_0N of (S), and R is the radius of circle (C) with center 0, therefore P also must be on the circle (C). Eventually the straight line (S) and the circle (C) have a common point P.

We already know that the circle (C) is self-symmetric in respect with the axis 00', thus the circle (C) and the straight line (S) have one more common point P' that is symmetrical with P in respect with point M_0.

Therefore, if the straight line (S) has one point inside and one point outside of a circle (C), they have two common points.

Comment: Despite of the logical demonstration above, the refuters still interpolate that this straight line and this circle have more than two common points.

To know this interpolation be true of false we suppose that the straight line (S) and the circle (C) intersect in some more points, for example P_1 and Q_1 are near by P and Q respectively.

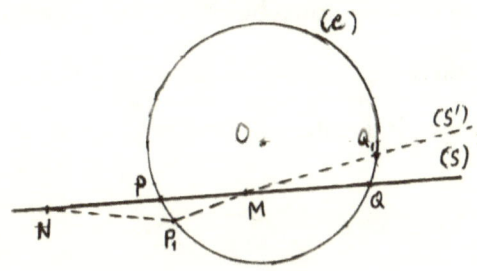

For this matter there are two case:

First Case: If P_1 and Q_1 locate on the circumference (C) that differ from P and Q respectively.

Firstly we join MP_1 and P_1N, we also get another straight line MP_1N that go through two given point MN. But according to postulate 1 of Euclid through MN there is only one straight line, therefore MP_1N and MPN must coincide one another.

Obviously P_1 cannot locate on the circumference (C) different from P. Therefore P_1 must be either at P or be on the straight line MN near by P.

Secondly, we join MQ_1 and get another straight line NMQ_1 produced, but NMQ is also a straight line, this is impossible because two straight lines cannot have a common portion NM, moreover the postulate 2 assumes that a straight line NM can be produced continuously in a straight line. So NMQ_1 and NMQ must coincide one another.

Obviously Q_1 cannot locate on the circumference (C)

different from Q. So Q_1 must be either at Q or be on the straight line NM produced.

Second case: P_1 locates on the straight line NM near by P.

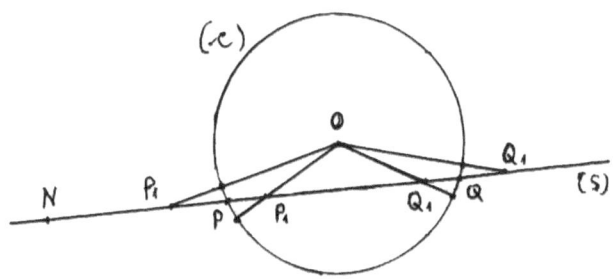

If P_1 locates on MN near by P, that means P_1 is inside or outside circle (C), because P is on the circumference (C).

If P_1 is inside circle (C) we have: $OP_1 < R$.

So P_1 cannot locate on the circle (C), eventually P_1 is not a common point of straight line (S) and circle (C).

If P_1 is outside circle (C) we have:

$OP_1 > R$

Therefore P_1 cannot locate an circle (C), eventually it is not a common point of straight line (S) and circle (C).

Same demonstration shows that Q_1 is not a common point of straight line (S) and circle (C).

S – *Conclusion*: The results above shows that a straight line has one point inside and one point outside a circle, it has two points common with the circle.

Problem 14: Two Circles Have Only Two Common Points

If in a given plane a circle (c) has one point x inside and one point y outside another circle (c'), the two circle intersects in two points.

Demonstration

Like problem 13, the demonstration must go through three cases below: refutation of the first group, demonstration of the second group, and demonstration of the author.

Refutation of the First Group

The first group of geometers interpolate that, in a given plane, a circle (c) has one point x inside and one point y outside another circle (c') having more than two common points.

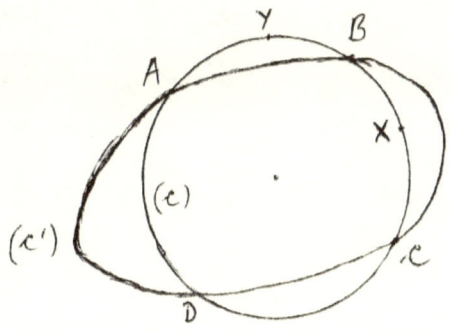

They refute this truth without respect to the definition and description of the circle to show two circles (c) and (c') having more than two points common. They describe them with deformation looking like the ellipse or the like. Moreover, they have no proof to justify their criticism. So we can ignore it. Therefore, the assumption that says "a circle (c) has one point x inside and one point y outside another circle (c') having utmost two points common" is still dependable.

Demonstration of the Second Group

They can likewise use the postulate of Dedekind to prove that, if in a given plane a circle (c) has one point x inside and one point y outside another circle (c'), the two circles intersect in two points. They prove the following:

Lemma

If O and O' are the centers of two circles (c) and (c') and R and R' are their radii, respectively, the straight line OO' meets the circle (c) at two points A and B, one of which is inside (c') and the other outside it.

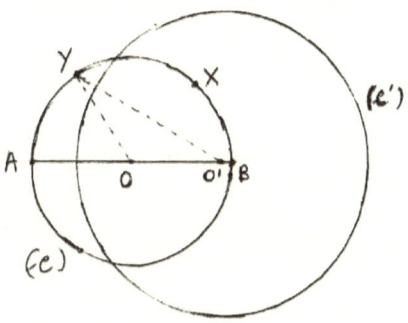

Now one of these points must fall (1) on the prolongation of OO' beyond O, (2) on OO' itself, or (3) on the prolongation of OO' beyond O'.

1. Suppose A lies on OO' produced. Then:
    $$AO' = AO + OO' = R + OO' \ldots (\alpha)$$

    But in the triangle OO'Y: O'Y < OY + OO. And because O'Y > R', OY = R. and R' < R + OO', it follows from (α) that AO' > R', and A, therefore, lies outside (c').

2. Suppose A lies on OO'. Then:
    $$OO' = OA + AO' = R + AO' \ldots (\beta)$$

    From the triangle OO'X, we have:
    $$OO' < OX + O'X.$$

    And because OX = R, O'X < R', it follows that:
    $$OO' < R + R'$$

    Hence, by (β), AO' < R', so A lies inside (c').

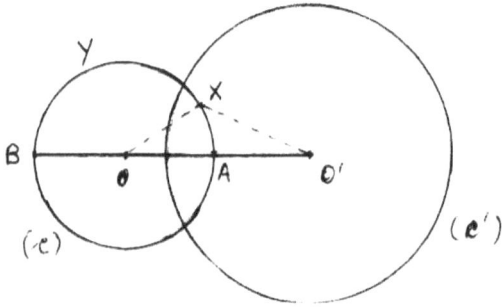

3. Suppose A lies on OO' produced. Then:
   $R = OA = OO' + O'A \ldots (\gamma)$

   And in the triangle OO'X: OX < OO' + O'X. That is R < OO' + O'X. Hence, by (γ): OO'+O'A < OO' + O'X. Or O'A < O'X, so A lies inside (c').

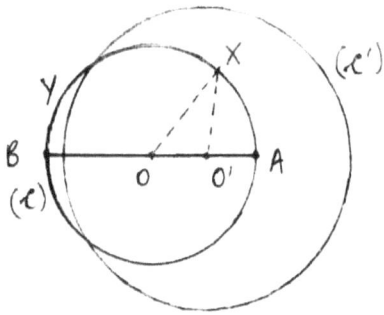

It is to be observed that one of the two points, A or B, is in the position of case (1), and the other is in the position of either case (2) or case (3); we must conclude that one of the two points, A or B, is inside and the other is outside the circle (c').

Proof of Theorem

The circle (c) is divided by the points A and B into two semicircles. Consider one of them, and suppose it to be described by a point moving from A to B. Take two separate points, P and Q, on it, and to fix our ideas, suppose that P precedes Q. Comparing the triangles OO'P and OO'Q, we observe that one side OO' is common, OP is equal to OQ, and the angle POO' is less than the angle QOO'. Therefore: O'P < O'Q.

Now consider the semicircle APQB is divided into two

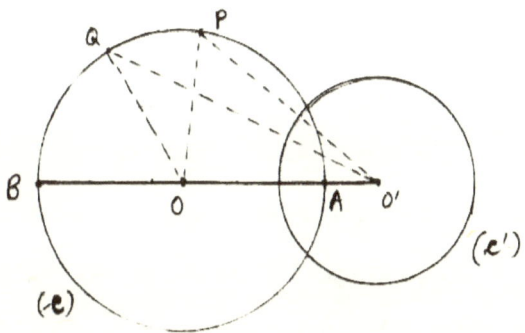

parts so the points of the first part are inside the circle (c') and those of the second part on the circumference of (c') are outside it. We have the conditions necessary for the applicability of the postulate of Dedekind (which is true for arcs of circles as for straight lines); there exists a point M separating the two parts. I say that O'M = R'. For if not, suppose O'M < R'. If then σ signifies the difference between R' and O'M, suppose a point M', which follows M, taken on the semicircle such that the chord MM' is not greater than σ. (For a way of doing this, see below.)

Then in the triangle O'MM': O'M' < O'M + MM' < O'M + σ. And therefore, O'M' < R'. If follow M', a point on the arc MB, is inside the circle (c'), which is absurd. Similarly, it may be proved that O'M is not greater than R. Hence, O'M = R. To find a point M' such that the chord MM' is not greater than σ, we may proceed thus.

Draw from M a straight line MP distinct from OM, and cut off MP on it equal to $\sigma/2$. Join OP, and draw another radius OQ such that the angle POQ is equal to the angle MOP. The intersection M' of OQ with the circle satisfies the required condition, for MM' meets OP at right angle in S.

Therefore, in the right angle triangle MSP, MS is not greater than MP. (It is less unless MP coincides with MS when it is equal.) Therefore, MS is not greater than $\sigma/2$, so MM' is not greater than σ.

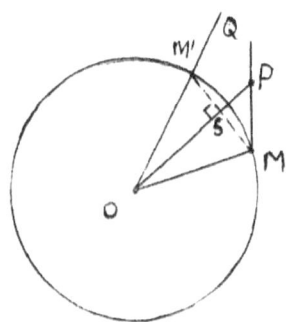

Comment

Considering this demonstration, at lemma (1) we find out a fact. But in the triangle OO'Y: O'Y < OY + OO'. This truth comes from Proposition 1.20, as we recognize in the problem 13 that it cannot exist without Proposition 1.1. But Proposition 1.1 assumes that two circles intersect in two points.

It's the same thing at lemma (2) and (3). Without Proposition 1.1, the commentators are not entitled to base on Proposition 1.20 for their demonstration. Finally, at the proof of theorem, there is one more fact. Comparing the triangle OO'P, OO'Q, we observe that on side OO' is common, OP is equal to OQ, and the angle POO' is less than the angle QOO'. Therefore, O'P < O'Q. This proof comes from Proposition 1.24.

But to prove Proposition 1.24, Euclid used Proposition 1.23, 1.19, and 1.5, respectively. Proposition 1.23 is deduced from Proposition 1.22 and 1.8. Proposition 1.22 is deduced from Proposition 1.20 and 1.3. But as we already recognize, Proposition 1.20 and 1.3 cannot exist without Proposition 1.1. And again, without Proposition 1.1 the commentators are not entitled to use Proposition 1.24 for their demonstration.

Conclusion

Obviously, the commentators use the assumption "two circles intersect in two points" to prove the same thing, "two circles intersect in two points." This is a confusion, so their demonstration is in vain.

Remark

Concerning the demonstration of second group, we must include that of Euclid. To prove his Proposition 3.10, he performs the following things. For, if possible, let the circle ABC cut the circle DEF at more than two, namely B, G, F, and H. Let BH and BG be joined and bisected at the points K and L, and from K and L, let KC and LM be drawn at right angles to BH and BG and carried through to the points A and E. This performance is of course based on Proposition 1.10 and 1.11, which relates strictly to Proposition 1.1. But Proposition 1.1 cannot exist without the assumption of "two circles intersect at two points" as a truth.

Demonstration of the Author

Like problem 13, we avoid to use the premised propositions that relate to two given circles that intersect in two points.

(A) – Firstly, we draw figure (1), in this figure, circle (C) is drawn at center 0 with radius R, circle (C') at center 0' with radius R'.

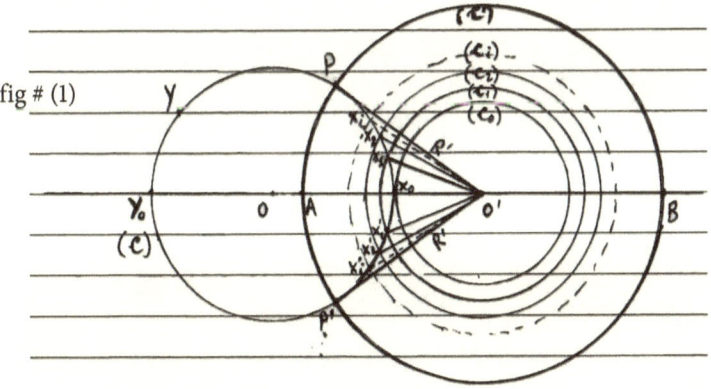

fig # (1)

Circle (C) cuts 00' produced at X_0 inside and Y_0 outside circle (C'). Let's the circle (C_0) be drawn at center 0' with radius O'X_0, it touches circle (C) at X_0, so the entire circle (C) is outside the circle (C_0), except the touching point X_0.

Due to the complicated figure (1), so the demonstration starts from the point X inside the circle (C') first then to the point Y outside (C') later.

1. Since the entire circle (C) is outside circle (C_o), so the portion of circumference (C) inside (C') is also outside (C_o). If X from X_o moves to X_1 we have:

$O'X_1 > O'X_o$ [X_1 is outside circle (C_o)]

2. Let's the circle (C_1) be drawn at center 0' with the radius $O'X_1$, since all the circles in figure (1) take the straight line 00' produced as their symmetric axis, so X_1 has its correspondent point X_1' in respect with that axis. Since X_o is inside circle (C_1) so the arc $X_1X_oX'_1$ of circle (C) is also inside the circle (C_1). Therefore the remaining arcs X_1P and X'_1P' are outside the circle (C_1).

If X moves from X_1 to X_2 we have:

$O'X_2 > O'X_1$ [X_2 is outside (C_1)]

3. Describe the circle (C_2) at center 0' with the radius $O'X_2$, the arc $X_2X_oX_2'$ is inside circle (C_2), so the remaining arcs X_2P and $X_2'P'$ are outside circle (C_2).

If X moves from X_2 to any points Xi between X_2 and P inside circle (C'), with the same demonstrations we have:

$R' > O'Xi > .... > O'X_2 > O'X_1 > O'X_o$ [R' is radius of (C')].

(B) Secondly, we draw the figure (2) as below:

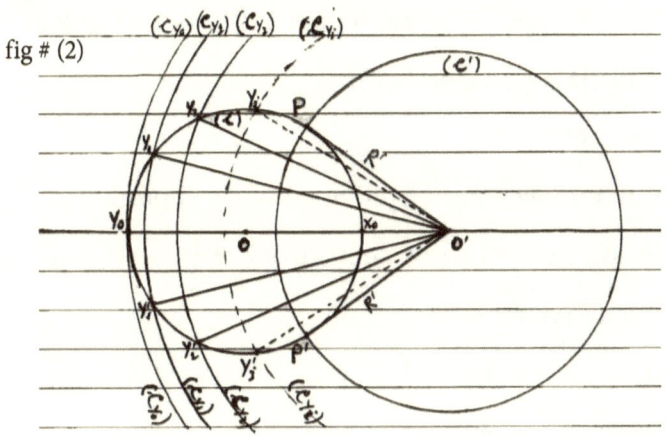
fig # (2)

In figure (2), if we describe a circle (C_{yo}) at center 0' with radius $O'Y_0$, it touches the circle (C) at Y_0. So the entire circle (C) is inside circle (C_{yo}), except the point Y_0.

1. If a point Y moves from Y_0 to Y_1, we describe a circle (C_{y1}) at center 0' with radius $O'Y_1$, since Y_1 is on the circle (C) so Y_1 is inside circle (C_{yo}).

Therefore we have:

$O'Y_0 > O'Y_1$ (the fourth property of a circle seen at problem 13)

2. We recognize that the arc $Y_1Y_0Y_1'$ of circle (C) is outside circle (C_{y1}), so the remaining two arcs Y_1P and $Y_1'P'$ are inside circle (C_{y1}).

If Y moves from Y_1 to Y_2 and we describe a circle (C_{y2}) at center 0' with radius $O'Y_2$, this circle is inside circle (C_{y1}).

Since Y_2 and Y_2' locate on arc Y_1P and $Y_1'P'$ respectively inside circle (C_{y1}), so we have:

$O'y_1 > O'Y_2$

3. If Y moves from Y_2 to any points Yi on the arc Y_0P of (C) inside circle (C_{y2}) but outside circle (C'), with the same demonstration we have:

$O'Y_2 > ... > O'Yi > R'$

Therefore we obtain all the results from Y_0 to Y_i:

$O'Y_0 > O'Y_1 > O'Y_2 > ... > O'Yi > R'$

4. Finally, summarize all the results from Y_0 to X_0 we obtain:

$O'Y_0 > O'Y_1 > O'Y_2 > ... > O'Yi > R' > O'Xi > ,,, > O'X_2 > O'X_1 > O'X_0$.

To fix our attention, the moving point from X_0 to Y_0 is called M instead X or Y.

The results above prove that a point M in connection with 0' when moves continuously from X_0 inside circle (C') to Y_0 outside (C'), it produces an infinite radii of circles that continuously increase from a minimum $O'X_0$ to a maximum $O'Y_0$ as below:

$O'Y_0 > O'Y_1 > O'Y_2 > ... > O'Yi > R' > O'Xi > ... > O'X_2 > O'X_1 > O'X_0$.

These series of radii are divided into two parts by the radius

R' of circle (C').

All the radii on the left of R' belong to the first part.

All the radii on the right of R' belong to the second part.

Since every radii of the first part precede R' and every radii of the second part follow R', therefore by the postulate of DedeKind, there must exist a radius O'P such that O'P = R'.

But we recognize that P is one of the positions on circle (C) that the moving point M goes through from X_o to Y_o, and O'P = R', thus P also must be on the circle (C').

This proves P be the common point of circle (C) and circle (C').

We already know that the circle (C) and (C') are self symmetric in respect to the straight line 00' produced as the symmetric axis. So the circle (C) and (C') have one more common point, that is the symmetric point P' of P in respect with straight line 00' produced.

Therefore, if a circle (C) has one point X inside and one point Y outside another circle (C'), the two circle intersect in two points.

Remark: In this demonstration, we focus only the case that the center 0 of circle (C) is outside the circle (C') in complying with the condition R'+R > OO'.

In case the center 0 of circle (C) is inside the circle (C') complying with the condition R'-R < 00', we can perform the same demonstration to obtain the same result.

(C). To deal with the criticism of the first group, we use the proposition 7 in book 1 of Euclid as an irrefutable proof to show that a circle (C) has one point X inside and one point Y outside another circle (C'), the two circle intersect in two point utmost.

The prop. 1.7 states as below:

"Given two straight lines constructed on the straight line (from its extremities) and meeting in a point, there cannot be constructed on the same straight line (from its extremities) and on the same side of it, two other straight lines meeting in another point and equal to the former two respectively, namely each to that which has the same extremities with it."

We recognize that:

-Prop. 1.7 relates to prop. 1.5

-Prop. 1.5 relates to prop. 1.4 and to the second case of prop. 1.3.

But prop. 1.4 is independent from prop. 1.1, and the second case of prop. 1.3 has been proved in the problem 13, it does not relate to prop. 1.1.

Therefore prop. 1.7 does not relate to prop. 1.1.

Now we come back to our matter:

According to the demonstration of the author, it's logical and clear to recognize that the two circles aforesaid intersect in two points only.

But the refuters interpolate that these two circles intersect in more than two points.

Let's 0 and 0' be the centers and R and R' be the radii of circle (C) and circle (C') respectively. We suppose two circles (C) and (C') having one more common point P'. Suppose P' and P being on the same upper side of 00' as the figure (3) show:

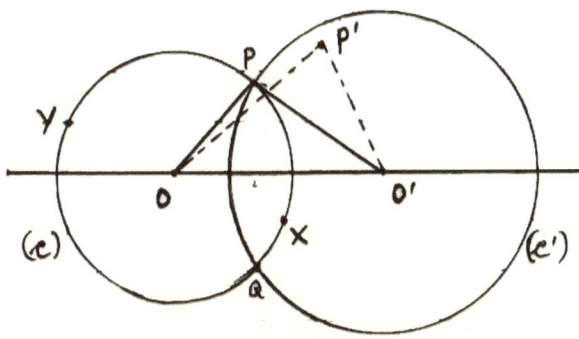

Joint the following points:

00', OP, O'P, OP' and O'P'.

Since P and P' are common points of two circles (C) and (C') so we have:

OP = OP' = R

O'P = O'P' = R'

We recognize that two straight lines OP and O'P are constructed on a straight line 00' (from its extremities) and meet in a point P.

According to the proposition 7 in book 1, there cannot be constructed on the same straight line 00' (from its extremities) and on the same side of it, two other straight lines meeting in another point and equal to the former two respectively.

That means P' cannot exist or no point P' which differ from P.

Same demonstration for the point Q' and Q on the lower side of 00' shows that Q' cannot exist.

Therefore we can conclude that two circles (C) and (C') intersect in two points utmost.

Part 2

Partial Permutations

Contents

Chapter 1 The Partial Permutations Introduction 145

Chapter 2 Solution of $P_n \binom{s}{t}$ 152

Chapter 3 The Permutation Rules 160

Chapter 4 The Distinction of Subpartial Permutations 170

Chapter 5 Relationship between $P_n \binom{s}{t}$ and Its Subpartial Permutations 179

Chapter 6 Solution Of $P_n \binom{o}{n}$ 205

Chapter I

The Partial Permutations Introduction

Preliminary: What Is Partial Permutations?

We already know that the permutations of n elements or n objects is nPn = n! If we set an order and a position for each element or object prior to perform the permutations nPn, we have n positions for n elements of nPn. After then, we will encounter many groups of permutations.

For example, considering the permutations 4P4 of four elements, e_1 e_2 e_3 e_4, in this order and at the original positions, E_1 E_2 E_3 E_4, respectively, then we permute these four elements as following.

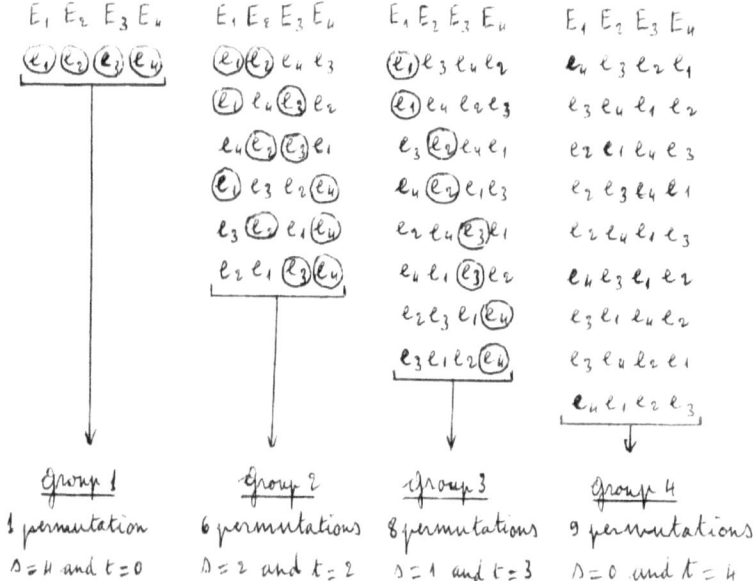

In this diagram, the encircled elements are stationed at their original positions. The others are transposed. The letter s represents the numbers of stationed elements; the letter t represents that of transposed elements.

I.I: Definition

The diagram above shows that the general permutations 4P4 is split into four groups of specific permutations, which I recommend to define as the partial permutations of four elements. In general, the general permutations nPn = n! consist of n groups of partial permutations of n elements.

I.2: Symbols Introduction

Prior to developing this theory, we must set the symbols for the partial permutations. In the diagram aforesaid, we recognize that each group of partial permutations possesses two specific properties:

- s are the numbers of stationed elements at their original positions; s are also the positive integers in the interval $0 \leq s \leq n$.
- t are the numbers of transposed elements; they are also the positive integers, but $t \neq 1$ because we cannot perform the transposition with only one element.

Therefore: $t = 0, 2, 3, 4 \dots n$. Based on these two properties, I recommend to introduce the symbols for the partial permutations of n elements as below:

$$P_n \binom{s}{t}$$

The symbol represents the partial permutations of n elements with s stationed elements at the top and t transposed elements at the bottom of the parentheses on the righthand side of P.

Important Remarks

- In $P_n\binom{s}{t}$, we always get a relationship $s + t = n$. It leads to $s = n - t$ or $t = n - s$. Therefore in some cases, we can write:

$$P_n\binom{s}{t} = P_n\binom{n-t}{t} \text{ or } P_n\binom{s}{t} = P_n\binom{s}{n-s}$$

Whichever cases, we have to keep in mind that the numbers of stationed elements are set at the top and those of the transposed elements are set at the bottom of the parentheses.

- In case of $s = 0$, we get $t = n$, it leads to $P_n\binom{0}{n}$ = unknown. Finding the permutations in the $P_n\binom{0}{n}$ is the master key of this theory.

- In case of $s = n$, we get $t = 0$, it leads to $P_n\binom{n}{0} = 1$. (See group 1 of the diagram aforesaid.)

- All the partial permutations $P_1\binom{0}{1}$, $P_2\binom{1}{1}$, $P_3\binom{2}{1}$ ... $P_n\binom{n-1}{1}$ actually do not exist because $t \neq 1$. However, for the latter demonstration purpose, we consider them as 0.

I.3: The Grade of $P_n\binom{s}{t}$

In $P_n\binom{s}{t}$, the numbers n are the positive integers in the interval $0 \leq n < +\infty$. Therefore, we define the grade of partial permutations $P_n\binom{s}{t}$ as following:

- The partial permutation of n = 0 is called zero-grade partial permutations. It is $P_0\binom{0}{0}=1$. (It will be proved later.)
- Any partial permutations of n = 1 are called first-grade partial permutations, such as $P_1\binom{1}{0}=1$ and $P_1\binom{0}{1}=0$.
- Any partial permutations of n = 2 are called second-grade partial permutations, such as $P_2\binom{2}{0}=1$, $P_2\binom{0}{2}=1$, and $P_2\binom{1}{1}=0$.
- For n = 3, 4, 5 ... n, they are called third-grade, fourth-grade, fifth-grade ... n^{th}-grade partial permutations, respectively.

I.4: The Expansions

There are two species expansions as below:

- The expansions of the general permutations nPn = n!
- The expansions of the partial permutations $P_n\binom{s}{t}$

The Expansions of nPn

The general permutations nPn can be split into n groups of $P_n\binom{s}{t}$. All of these groups are the expansions of nPn. Let's expand the general permutations from 0P0 = 0! to nPn = n!:

- $0! = P_0 \binom{0}{0}$
- $1! = P_1 \binom{1}{0}$
- $2! = P_2 \binom{2}{0} + P_2 \binom{0}{2}$
- $3! = P_3 \binom{3}{0} + P_3 \binom{1}{2} + P_3 \binom{0}{3}$
- $4! = P_4 \binom{4}{0} + P_4 \binom{2}{2} + P_4 \binom{1}{3} + P_4 \binom{0}{4}$

...

- $n! = P_n \binom{n}{0} + P_n \binom{n-2}{2} + P_n \binom{n-3}{3} + \ldots + P_n \binom{1}{n-1} + P_n \binom{0}{n}$

The Expansions of $P_n \binom{s}{t}$

The partial permutations $P_n \binom{s}{t}$ consist of a numbers of permutations that are specified by s stationed elements and t transposed elements. All these specific permutations are the expansions of $P_n \binom{s}{t}$. Consider the circled e_i as the stationed elements and the uncircled e_i as the transposed elements. Let's expand some low-grade partial permutations:

- $*P_1 \binom{1}{0} = [e_1] = 1$ permutation

- $*P_2 \binom{2}{0} = [e_1 e_2] = 1 \quad P_2 \binom{0}{2} = [e_2 e_1] = 1$

- $*P_3 \binom{3}{0} = [e_1 e_2 e_3] = 1 \quad P_3 \binom{1}{2} = \begin{bmatrix} e_1 e_3 e_2 \\ e_3 e_2 e_1 \\ e_2 e_1 e_3 \end{bmatrix} = 3 \quad P_3 \binom{0}{3} = \begin{bmatrix} e_2 e_3 e_1 \\ e_3 e_1 e_2 \end{bmatrix} = 2$

- $*P_4\binom{4}{0} = [e_1e_2e_3e_4] = 1$ $\quad P_4\binom{2}{2} = \begin{bmatrix} e_1e_2e_4e_3 \\ e_1e_4e_3e_2 \\ e_4e_2e_3e_1 \\ e_1e_3e_2e_4 \\ e_3e_2e_1e_4 \\ e_2e_1e_3e_4 \end{bmatrix} = 6$

- $*P_4\binom{1}{3} = \begin{bmatrix} e_1e_3e_4e_2 \\ e_1e_4e_2e_3 \\ e_3e_2e_4e_1 \\ e_4e_2e_1e_3 \\ e_2e_4e_3e_1 \\ e_4e_1e_3e_2 \\ e_2e_3e_1e_4 \\ e_3e_1e_2e_4 \end{bmatrix} = 8$ $\quad P_4\binom{0}{4} = \begin{bmatrix} e_4e_3e_2e_1 \\ e_3e_4e_1e_2 \\ e_2e_1e_4e_3 \\ e_2e_3e_4e_1 \\ e_2e_4e_1e_3 \\ e_4e_3e_1e_2 \\ e_3e_1e_4e_2 \\ e_3e_4e_2e_1 \\ e_4e_1e_2e_3 \end{bmatrix} = 9$

By experience, it's easy to expand the low-grade partial permutations, as we have done above. For the high-grade partial permutations $P_n\binom{s}{t}$, we have to take a lot of work and time to expand them. However, we can later find out the numbers of specific permutations contained in $P_n\binom{s}{t}$.

Chapter 2

Solution of $P_n \binom{s}{t}$

Preliminary

Our task is how to know the numbers of permutations in $P_n \binom{s}{t}$. So we must focus on the s stations elements of $P_n \binom{s}{t}$ to find out any relationship between s, t, and n.

2.1: The Role of s in $P_n \binom{s}{t}$

We first mention the combinations ^nC_s of n elements with s elements be selected at a time. The numbers of combinations of ^nC_s are given by formula:

$$^nC_s = \frac{n!}{(n-s)! s!}$$

Because the combinations disregard order, so a set of s combinatorial elements in ^nC_s such as (e_1 e_2 e_3 ... e_s) are considered one combination, not s! permutations. Now we consider the partial permutations $P_n \binom{s}{t}$ of n elements of which s elements are selected to stay at their original positions and connect with t transposed elements as the one of the permutations contained in $P_n \binom{s}{t}$:

$$\underbrace{\begin{array}{cccc} E_1 & E_2 & E_3 \ldots\ldots E_s \\ e_1 & e_2 & e_3 \ldots\ldots e_s \end{array}}_{\text{s stationed elements}} \underbrace{\begin{array}{cccc} E_{s+1} & E_{s+2} \ldots\ldots E_{n-1} & E_n \\ e_{s+4} & e_{s+7} \ldots\ldots e_n & e_{n-1} \end{array}}_{\text{t transposed elements}}.$$

For convenience, we create one special arrangement of s and t as the diagram describes above. This arrangement of s elements is only one way. We cannot get another way by transposing any two elements among s. If we do so, we will get another arrangement of s elements in which two elements are transposed from their original positions. But the partial permutations $P_n \binom{s}{t}$ specify that s are a set of stationed elements only, not a set of s − 2 of stationed elements. Therefore, the s in $P_n \binom{s}{t}$ are equivalent to the s in combinations ^nC_s.

2.2: Theorem I

Consider the partial permutations $P_n \binom{s}{t}$ if there are lower-grade partial permutations $P_t \binom{o}{t}$ of t elements and a combination ^nC_s of n element with s elements be selected at a time. The theorem states that the partial permutations $P_n \binom{s}{t}$ are equal to the product of ^nC_s and $P_t \binom{o}{t}$:

$$P_n \binom{s}{t} = {}^n C_s . P_t \binom{o}{t}$$

Demonstration

As we already know, s stationed elements in $P_n\binom{s}{t}$ are equivalent to s elements selected at a time of combination ^nC_s. Therefore, in $P_n\binom{s}{t}$, the numbers of combinations of n elements with s elements are selected to stay at their positions equal to ^nC_s. Each set of s stationed elements is one combination of ^nC_s and connects to t transposed elements, as the diagram describes:

$$\underbrace{E_1 \quad E_2 \quad E_3 ...... E_s}_{\text{s stationed elements}} \underbrace{E_{s+1} \quad E_{s+2} ...... E_{n-1} \quad E_n}_{} $$
$$\underbrace{e_4 \quad e_2 \quad e_3 .... e_s}_{\text{s stationed elements}} \underbrace{e_{s+3} \quad e_{s+5} ...... e_n \quad e_{n-3}}_{\text{t transposed elements}}.$$

For convenience, we create one special arrangement of s in connection with t transposed elements, as the diagram describes above. We recognize that the set $\left[e_1e_2e_3.....e_s\right]$ of s elements is one way only. It is equivalent to one combination of ^nC_s, while each t transposed element in the set $[e_{s+3}\,e_{s+5}\,...\,e_n\,e_{s-3}]$ of t elements can transpose to any other positions (except its original position). These transpositions produce many other sets of t transposed elements. But all these sets of t transposed elements are the expansions of $P_t\binom{o}{t}$ by definition of expansions of $P_n\binom{s}{t}$. Therefore, the combination $\left[e_1e_2e_3.....e_s\right]$ connects with the expansions of $P_t\binom{o}{t}$. But this combination is the one of the combinations ^nC_s. Hence, these connections produce the numbers of permutations [s & t], which are equal to the product $^nC_s \cdot P_t \binom{o}{t}$. But these permutations [s & t] are the expansions of $P_n \binom{s}{t}$ by definition of expansions of $P_n \binom{s}{t}$. Therefore, we can conclude:

$$P_n \binom{s}{t} = {}^n C_s \cdot P_t \binom{o}{t}.$$

Corollary I

In the formula:

$$P_n \binom{s}{t} = {}^n C_s \cdot P_t \binom{o}{t}$$

If s = n and t = 0, we get:

$$P_n \binom{n}{o} = {}^n C_n \cdot P_0 \binom{0}{0}$$

But $P_n \binom{n}{o} = 1$ and $^nC_n = 1$

Therefore, we deduce:

$$P_0 \binom{0}{0} = 1$$

The case makes us to be astonished; however, it is a truth and more logical than 0! = 1.

Corollary 2

We already know that the general permutation n! can be split or expanded into n groups of partial permutations as following:

$$n! = P_n \binom{n}{0} + P_n \binom{n-2}{2} + P_n \binom{n-3}{3} + \ldots + P_n \binom{3}{n-3} + P_n \binom{2}{n-2} + P_n \binom{1}{n-1} + P_n \binom{0}{n}.$$

If we expand consecutively all the down-grade permutations from n!, we get:

- $(n-1)! = P_{n-1} \binom{n-1}{0} + P_{n-1} \binom{n-3}{2} + P_{n-1} \binom{n-4}{3} + \ldots + P_{n-1} \binom{2}{n-3} + P_{n-1} \binom{1}{n-2} + P_{n-1} \binom{0}{n-1}$
- $(n-2)! = P_{n-2} \binom{n-2}{0} + P_{n-2} \binom{n-4}{2} + P_{n-2} \binom{n-5}{3} + \ldots + P_{n-2} \binom{1}{n-3} + P_{n-2} \binom{0}{n-2}$
- $(n-3)! = P_{n-3} \binom{n-3}{0} + P_{n-3} \binom{n-5}{2} + P_{n-3} \binom{n-6}{3} + \ldots + P_{n-3} \binom{1}{n-4} + P_{n-3} \binom{0}{n-3}$

....
....

- $4! = P_4 \binom{4}{0} + P_4 \binom{2}{2} + P_4 \binom{1}{3} + P_4 \binom{0}{4}$
- $3! = P_3 \binom{3}{0} + P_3 \binom{1}{2} + P_3 \binom{0}{3}$
- $2! = P_2 \binom{2}{0} + P_2 \binom{0}{2}$
- $1! = P_1 \binom{1}{0}$
- $0! = P_0 \binom{0}{0}$

Finally, we get the relationship:
$$P_0 \binom{0}{0} = 0!$$

But Corollary 1 gives us:
$$P_0 \binom{0}{0} = 1$$

Ancient and Modern Mathematics

Therefore, we deduce:

$0! = 1.$

So $0! = 1$ is not a convention but a corollary of Theorem 1.

Application

We can apply Theorem 1 to solve $P_n \binom{0}{n}$ from the low- to the higher-grade partial permutations as following:

- We know $P_2 \binom{0}{2} = 1$ by the expansions of $P_n \binom{s}{t}$ or the expansions of 2!.

  as:
  $$P_2 \binom{0}{2} = 2! - \left[P_2 \binom{2}{0} \right]$$
  $$\phantom{P_2 \binom{0}{2}} = 2 \ - \ 1$$

  Thus, we get: $P_2 \binom{0}{2} = 1$.

- With $P_2 \binom{0}{2} = 1$, we proceed to $P_3 \binom{0}{3}$ by applying Theorem 1 in connection with the expansion of general permutations 3!:

  $$P_3 \binom{0}{3} = 3! - \left[P_3 \binom{3}{0} + P_3 \binom{1}{2} \right]$$
  $$\phantom{P_3 \binom{0}{3}} = 6 \ - \left[\ 1 \ + {}^3C_1 . P_2 \binom{0}{2} \right] \rightarrow \text{(Theorem 1)}$$
  $$\phantom{P_3 \binom{0}{3}} = 6 \ - \ 1 \ - \ 3.P_2 \binom{0}{2}$$
  $$\phantom{P_3 \binom{0}{3}} = 6 \ - \ 1 \ - \ 3.1$$

Thus, we have:

$$P_3\binom{0}{3} = 2$$

With $P_3\binom{0}{3} = 2$, we proceed to $P_4\binom{0}{4}$ as:

$$P_4\binom{0}{4} = 4! - \left[P_4\binom{4}{0} + P_4\binom{2}{2} + P_4\binom{1}{3}\right]$$

$$= 24 - \left[1 + {}^4C_2 \cdot P_2\binom{0}{2} + {}^4C_1 \cdot P_3\binom{0}{3}\right] \rightarrow \text{(Theorem 1)}$$

$$= 24 \ - \ 1 \ - \ 6.1 \ - \ 4.2$$

Therefore, we have:

$$P_4\binom{0}{4} = 9$$

With $P_4\binom{0}{4} = 9$, we proceed to $P_5\binom{0}{5}$:

$$P_5\binom{0}{5} = 5! - \left[P_5\binom{5}{0} + P_5\binom{3}{2} + P_5\binom{2}{3} + P_5\binom{1}{4}\right]$$

$$= 120 - \left[1 + {}^5C_3 \cdot P_2\binom{0}{2} + {}^5C_2 \cdot P_3\binom{0}{3} + {}^5C_4 \cdot P_4\binom{0}{4}\right]$$

$$= 120 - \ 1 \ - \ 10.1 \ - \ 10.2 \ - \ 5.9$$

Therefore, we have:

$$P_5\binom{0}{5} = 44$$

For the high-grade partial permutations $P_n\binom{0}{n}$, we can use this method to proceed from $P_5\binom{0}{5}$ to $P_6\binom{0}{6}$ and then from $P_6\binom{0}{6}$ to $P_7\binom{0}{7}$ and so on. However, that method is not a good way because we must take a lot of time and work to solve the high-grade partial permutations $P_n\binom{0}{n}$. Therefore, we must find out a general formula applicable for any grade $P_n\binom{0}{n}$.

Chapter 3

The Permutation Rules

Introduction

The permutation rules are the performance method to determine the relationship between the partial permutations $P_n\binom{s}{t}$ and $P_{n-1}\binom{S}{T}$. These methods are described as following. Considering the partial permutations $P_{n-1}\binom{S}{T}$, if there is another stationed element e_n that joins to permute with $(n-1)$ elements of $P_{n-1}\binom{S}{T}$, it creates the subpartial permutations of $P_n\binom{s}{t}$ via the rules that we call the permutation rules. The subpartial permutations of $P_n\binom{s}{t}$ means a part of the partial permutations $P_n\binom{s}{t}$. Prior to defining and performing the permutation rules, we must recognize some notations and terminologies as following:

- e_n joins to transpose with S elements of $P_{n-1}\binom{S}{T}$: <u>S tr.</u>
- e_n joins to transpose with T elements of $P_{n-1}\binom{S}{T}$: <u>T tr.</u>
- e_n joins to partial permutations $P_{n-1}\binom{S}{T}$: <u>J</u>

- Subpartial permutations of $P_n\binom{s}{t}$ formed by e_n, which transposes with S elements of

  $P_{n-1}\binom{S}{T}: S^S P_n\binom{s}{t} \left[S^S P_n\binom{s}{t} < P_n\binom{s}{t} \right]$

- Subpartial permutations of $P_n\binom{s}{t}$ formed by e_n, which transposes with T elements of

  $P_{n-1}\binom{S}{T}: S^T P_n\binom{s}{t} \left[S^T P_n\binom{s}{t} < P_n\binom{s}{t} \right]$

- Subpartial permutations of $P_n\binom{s}{t}$ formed by e_n, which joins the $P_{n-1}\binom{S}{T}: S^J P_n\binom{s}{t} \left[S^J P_n\binom{s}{t} < P_n\binom{s}{t} \right]$

The correspondence of two elements is defined as below:

- Considering two single permutations of $P_n\binom{s}{t}$ and n positions, each single permutation has n elements of which each element stays at one of n positions. In these two single permutations, any two elements that stay at the same position are defined as two correspondent elements.

Example

Considering two single permutations (1) and (2) of five elements (2 & 3) and five positions $E_1\ E_2\ E_3\ E_4\ E_5$, we have:

$$\begin{array}{ccccc} E_1 & E_2 & E_3 & E_4 & E_5 \\ [e_1 & e_4 & e_3 & e_5 & e_2] \\ \updownarrow & \updownarrow & \updownarrow & \updownarrow & \updownarrow \\ [e_2 & e_4 & e_3 & e_1 & e_5] \end{array}$$ (1)

(2)

- e_1 of (1) and e_2 of (2) at E_1 are correspondent.
- e_4 of (1) and e_4 of (2) at E_2 are correspondent.

...

- e_2 of (1) and e_5 of (2) at E_5 are correspondent.

The order of an element is defined as:

- Considering n elements of $P_n \binom{s}{t}$ and n positions of these elements, to avoid the confusion, these elements and these positions must be set in dual orderly arrangements from first to n^{th} order as following:

- Positions: $E_1 \, E_2 \, E_3 \, \ldots \, E_{n-2} \, E_{n-1} \, E_n$ (stationary)
- Elements: $e_1 \, e_2 \, e_3 \, \ldots \, e_{n-2} \, e_{n-1} \, e_n$ (transposable)
- E_1 is the original position e_1.
- E_2 is the original position e_2.

...

- E_n is the original position e_n.

The order of element e is identified by the adjacent index on the righthand side of e $(e_1, e_2 \ldots e_n)$. Because the elements of $P_n \binom{s}{t}$ consist of s stationed and t transposed elements, so the order of element e_i is always unchanged even as they move away from their original positions.

The coincidence of two permutations is defined as:

- If two single permutations have exactly the same stationed elements and the same transposed elements in which any two correspondent elements of these permutations get the same order, then these two single permutations are coincident one another and considered as one single permutation.

Example

Consider four single permutations of five elements named (1), (2), (3), (4), and 5 positions as following:

$E_1\ E_2\ E_3\ E_4\ E_5$		$E_1\ E_2\ E_3\ E_4\ E_5$
$[e_1\ e_5\ e_3\ e_2\ e_4]$ (1)		$[e_1\ e_3\ e_5\ e_4\ e_2]$ (3)
Coincident		not coincident

| [e₁ e₅ e₃ e₂ e₄] (2) | | [e₂ e₄ e₃ e₁ e₅] (4) |

3.I: First Rule

Considering the partial permutations $P_{n-1}\binom{s}{T} = P_{n-1}\binom{s+1}{t-2}$, if another stationed element e_n joins to transpose with s + 1 stationed elements of $P_{n-1}\binom{s+1}{t-2}$, it forms the subpartial permutations $S^S P_n \binom{s}{t}$ of which the numbers of permutations are equal to those of $(s+1).P_{n-1}\binom{s+1}{t-2}$. The S transposition rule is described by the diagram:

$$P_{n-1}\binom{s+1}{t-2} \underline{\text{S.tr.}} e_n \underline{\text{S.tr.}} (s+1).P_{n-1}\binom{s+1}{t-2} = S^S P_n \binom{s}{t}$$

(S = s + 1, T = t − 2)

Why? Each permutation of $P_{n-1}\binom{s+1}{t-2}$ consists of s + 1 stationed elements and (t − 2) transposed elements in the relationship [s + 1 + (t − 2)] = n − 1. We can consider [s + 1 & (t − 2)] as one single permutation of $P_{n-1}\binom{s+1}{t-2}$. Let's join and transpose e_n with one of the elements s + 1. It causes s + 1 to lose one stationed element while (t − 2) gets two more transposed elements because that stationed element and e_n interchange each other their positions so they become two transposed elements.

Therefore, the transposition between e_n and one of the elements $s + 1$ in one single permutation $[s + 1 \, \& \, (t - 2)]$ forms one single permutation $[s \, \& \, t]$ for $S^S P_n \binom{s}{t}$: $[s + 1 \, \& \, (t - 2)] = [s \, \& \, t] = 1$ permutation.

But $[s + 1 \, \& \, (t - 2)]$ consists of $s + 1$ stationed elements. Thus, e_n must transpose $(s + 1)$ times to achieve the transpositions. Each time causes $s + 1$ to lose one stationed element and $(t - 2)$ to get two more transposed elements. Thus, it produces:

$$(s + 1) \times [s + 1 \, \& \, (t - 2)] = (s + 1) \times [s \, \& \, t] \quad (1)$$

If we multiply both members of (1) by $P_{n-1} \binom{s+1}{t-2}$, we get:

$$(s+1).\left[s+1 \, \& \, (t-2)\right].P_{n-1}\binom{s+1}{t-2} = (s+1).\left[s \, \& \, t\right].P_{n-1}\binom{s+1}{t-2}$$

We can rearrange:

$$(s+1).P_{n-1}\binom{s+1}{t-2}\left[s+1 \, \& \, (t-2)\right] = (s+1).P_{n-1}\binom{s+1}{t-2}\left[s \, \& \, t\right]$$

But $(s+1).P_{n-1}\binom{s+1}{t-2}\left[s \, \& \, t\right]$ are the numbers of permutations specified by s stationed elements and t transposed elements. Those numbers are called subpartial permutations created by S transposition:

$$S^S P_n \binom{s}{t}, \text{ a part of } P_n \binom{s}{t}$$

Therefore, we can write:

$$S^S P_n \binom{s}{t} = (s+1).P_{n-1}\binom{s+1}{t-2}\left[s+1 \, \& \, (t-2)\right]$$

But $[s + 1 \ \& \ (t - 2)] = 1$ (one of the permutations of $P_{n-1}\binom{s+1}{t-2}$).
Thus, we have the first rule:

$$S^S P_n \binom{s}{t} = (s+1).P_{n-1}\binom{s+1}{t-2}$$ (the case requires: $n \geq 2$ and $t \geq 2$)

3.2: Second Rule

Considering the partial permutations $P_{n-1}\binom{s}{T} = P_{n-1}\binom{s}{t-1}$, if another stationed element e_n joins to transpose with $(t - 1)$ transposed elements of $P_{n-1}\binom{s}{t-1}$, it forms the subpartial permutations $S^T P_n \binom{s}{t}$ of which the numbers of permutations are equal to those of $(t-1).P_{n-1}\binom{s}{t-1}$. The T transposition rule is described by the diagram:

$$P_{n-1}\binom{s}{t-1} \ \underline{T.tr.} \ e_n \ \underline{T.tr.} \ (t-1).P_{n-1}\binom{s}{t-1} = S^T P_n \binom{s}{t}$$

$(T = s - 1 \text{ and } S = s)$

Why? Like the first rule, each permutation of $P_{n-1}\binom{s}{t-1}$ is considered as $[s \ \& \ (t - 1)]$. If another stationed element e_n joins to transpose with one of the $(t - 1)$ transposed elements in $[s \ \& \ (t - 1)]$, it causes $(t - 1)$ transposed elements to get one more transposed element e_n because e_n transposes to the other position. Thus, we get $[s \ \& \ t]$. This is one of the permutations of $S^T P_n \binom{s}{t}$.

The above shows that each transposition between e_n and one element of (t − 1) in one single permutation [s & (t − 1)] produces one single permutation [s & t] for $S^T P_n \binom{s}{t}$. But [s & (t − 1)] consists of (t − 1) transposed elements, so e_n must transpose (t − 1) times to achieve the transpositions. Therefore, it produces (t − 1) x [s & (t − 1)] or (t − 1) x [s & t] permutations for $S^T P_n \binom{s}{t}$: (t − 1) x [s & (t − 1)] = (t − 1) x [s & t] (1). If we multiply both members of (1) by $P_{n-1} \binom{s}{t-1}$, we get:

$$(t-1).\left[s \& (t-1)\right].P_{n-1}\binom{s}{t-1} = (t-1).\left[s \& t\right].P_{n-1}\binom{s}{t-1}$$

If we rearrange this relationship, we get:

$$(t-1).P_{n-1}\binom{s}{t-1}\left[s \& (t-1)\right] = (t-1).P_{n-1}\binom{s}{t-1}\left[s \& t\right]$$

But $(t-1).P_{n-1}\binom{s}{t-1}\left[s \& t\right]$ are the numbers of permutations specified by s stationed elements and t transposed elements. Those numbers are called subpartial permutations created by T transposition: $S^T P_n \binom{s}{t}$ [another part of $P_n \binom{s}{t}$]. Therefore, we can write:

$$S^T P_n \binom{s}{t} = (t-1).P_{n-1}\binom{s}{t-1}\left[s \& (t-1)\right]$$

But [s & (t − 1)] = 1 (one of the permutations of $P_{n-1}\binom{s}{t-1}$). Hence, we have the second rule:

$$S^T P_n \binom{s}{t} = (t-1).P_{n-1}\binom{s}{t-1} \text{ (the case requires: } n \geq 3 \text{ and } t \geq 3\text{)}.$$

3.3: Third Rule

Considering the partial permutations $P_{n-1}\binom{S}{T} = P_{n-1}\binom{s-1}{t}$, if another stationed element e_n joins the partial permutations $P_{n-1}\binom{s-1}{t}$ with no transposition, it forms the subpartial permutations $S^J P_n \binom{s}{t}$ of which the numbers of permutations are equal to those of $P_{n-1}\binom{s-1}{t}$. This join rule is described by the diagram:

$$P_{n-1}\binom{s-1}{t} \underline{J} e_n \overrightarrow{J} P_{n-1}\binom{s-1}{t} = S^J P_n \binom{s}{t} \quad (T = t,\ S = (s-1))$$

Why? Like the other rules, we consider [s − 1 & t] as one single permutation of $P_{n-1}\binom{s-1}{t}$, and let's e_n join [s − 1 & t] without transposition. That means the joining adds e_n to s − 1 of [s − 1 & t]. Thus, we get [s & t]. This is one single permutation of $S^J P_n \binom{s}{t}$. The above shows that, for one single permutation of $P_{n-1}\binom{s-1}{t}$, we get one single permutation of $S^J P_n \binom{s}{t}$. We can write: [s − 1 & t] = [s & t] (one permutation). If we multiply both members by $P_{n-1}\binom{s-1}{t}$, we get:

$$[s-1\ \&\ t].P_{n-1}\binom{s-1}{t} = [s\ \&\ t].P_{n-1}\binom{s-1}{t}$$

Or we can rearrange:

$$P_{n-1}\binom{s-1}{t}[s\ \&\ t] = P_{n-1}\binom{s-1}{t}[s-1\ \&\ t] \quad (1)$$

But $P_{n-1}\binom{s-1}{t}[s\,\&\,t]$ are the numbers permutations specified by s stationed elements and t transposed elements. Those numbers are called subpartial permutations created by join permutation: $S^J P_n\binom{s}{t}$ [another part of $P_n\binom{s}{t}$]. Therefore, we can write (1) as:

$$S^J P_n\binom{s}{t} = P_{n-1}\binom{s-1}{t}[s-1\,\&\,t]$$

But [s − 1 & t] is one single permutation of $P_{n-1}\binom{s-1}{t}$. Thus, we get the third rule:

$$S^J P_n\binom{s}{t} = P_{n-1}\binom{s-1}{t} \text{ (the case requires } n \geq 1, s \geq 1, \text{ and } t \neq 1).$$

Chapter 4

The Distinction of Subpartial Permutations

Preliminary

We have recognized three species of subpartial permutations as following:

- $S^S P_n \binom{s}{t} = (s+1) \cdot P_{n-1} \binom{s+1}{t-2} \cdot [s \& t]$ (first rule)
- $S^T P_n \binom{s}{t} = (t-1) \cdot P_{n-1} \binom{s}{t-1} \cdot [s \& t]$ (second rule)
- $S^J P_n \binom{s}{t} = P_{n-1} \binom{s-1}{t} \cdot [s \& t]$ (third rule)

All consist of a numbers of single permutation [s & t]. But among these numbers of single permutation, there are no coincidences of one another. To prove this, we must denote the properties of three subpartial permutations above.

4.I: The Properties of $S^S P_n \binom{s}{t}$

Based on the first rule, the single permutation [s & t] of $S^S P_n \binom{s}{t}$ is created by e_n, which joins to transpose with one stationed element of s + 1 in the single permutation [s + 1 & (t − 2)] of $P_{n-1} \binom{s-1}{t-2}$. That means e_n from its original position E_n moves to the original position of the stationed element aforesaid, and this stationed element moves to the original position E_n of e_n. So these two stationed elements become two transposed elements of single permutation $[s \& t] \in S^S P_n \binom{s}{t}$.

Let's take an example that e_n transposed with e_{s+1} in a special single permutation $[s+1 \& (t-2)] \in P_{n-1}\binom{s+1}{t-2}$ creates a special single permutation $[s \& t] \in S^S P_n \binom{s}{t}$ as Diagram (1) describes.

$$
\begin{array}{c}
E_1 \quad E_2 \quad E_3 \ldots E_{n-1} \quad E_n \quad E_{n+1} \quad E_{n+2} \quad E_{n+3} \ldots E_{n-3} \quad E_{n-2} \quad E_{n-1} \quad E_n \\
\downarrow \downarrow \downarrow \quad \downarrow \downarrow \downarrow \quad \downarrow \quad \downarrow \quad \downarrow \quad \downarrow \quad \downarrow \quad \downarrow
\end{array}
$$

| e_1 e_2 e_3 ... e_{s-1} e_s e_{s+1} e_{n-2} e_{n-1} ... e_{s+4} e_{s+3} e_{s+2} e_n |
| $[s+1] \& (t-2) \in P_{n-1}\binom{s+1}{t-2}$ S &t. |
| e_1 e_2 e_3 ... e_{s-1} e_s e_n e_{n-2} e_{n-1} ... e_{s+4} e_{s+3} e_{s+2} e_{s+1} |
| $[s \& t] \in S^S P_n \binom{s}{t}$ |

Diagram (1) shows that the single permutation [s & t] of $S^S P_n \binom{s}{t}$ has the last element e_n that locates at the position E_{s+1} and e_{s+1} locates at the last position E_n.

171

In the example above, if e_n transposes with e_s or e_{s-1} ..., we get the same results. The last element e_n locates at position E_s or E_{s-1} ... and e_s, or e_{s-1} ... locates at the last position E_n. But [s & t] is one of the permutations of $S^S P_n \binom{s}{t}$ that equals to $(s+1).P_{n-1}\binom{s+1}{t-2}[s \& t]$. If we perform the S transposition between e_n and s+1 stationed elements in all single permutations [s+1 & (t − 2)] of $P_{n-1}\binom{s+1}{t-2}$, we get the whole single permutations [s & t] of $S^S P_n \binom{s}{t}$ with the same results above. Therefore, we deduce that the subpartial permutations $S^S P_n \binom{s}{t}$ has two properties as following:

- In any single permutation of $S^S P_n \binom{s}{t}$, there is always a transposed element that locates at the last position E_n.
- In any single permutation of $S^S P_n \binom{s}{t}$, the last element e_n always locates at the original position of the transposed element, which is locating at last position E_n

4.2: The Properties of $S^T P_n \binom{s}{t}$

Based on the second rule, the single permutation [s & t] of $S^T P_n \binom{s}{t}$ is created by a stationed element e_n, which joins to transpose with one transposed element of (t − 1) in the single permutation [s & (t − 1)] of $P_{n-1}\binom{s}{t-1}$. That means e_n, from its original position E_n, moves to the current position (not original) of the transposed element aforesaid. This transposed element moves to the original position E_n of e_n. So e_n and this transposed element become two transposed elements in a single permutation [s & t].

Let's take an example that e_n transposes with a transposed element e_{s+1} at the current position E_{n-1} in a special single permutation of $P_{n-1}\binom{s}{t-1}$, as Diagram (2) describes:

$$E_1 \quad E_2 \quad E_3 \ldots E_{n-1} \quad E_n \quad E_{n+1} \quad E_{n+2} \quad E_{n+3} \ldots E_{n-3} \quad E_{n-2} \quad E_{n-1} \quad E_n$$

$$\boxed{e_1 \;\; e_2 \;\; e_3 \ldots e_{n-1} \;\; e_n \;\; \ell_{n-3} \;\ell_{n-2}\; \ell_{n-1} \ldots \ell_{n+3}\; \ell_{n+2}\; \ell_{n+1}} \quad \boxed{e_n}$$
$[\text{\textcircled{A}} \,\&\, (t-1)] \in P_{n-1}\binom{s}{t-1}$ \hfill T.h.

$$\boxed{e_1 \;\; e_2 \;\; e_3 \ldots e_{n-1} \;\; e_n \;\; \ell_{n-3}\; \ell_{n-2}\; \ell_{n-1} \ldots \ell_{n+3}\; \ell_{n+2}\; \overset{x}{e_n} \;\; \overset{y}{\ell_{n+1}}}$$
$[\text{\textcircled{A}} \,\&\, t] \in S^T P_n\binom{s}{t}$

Diagram (2) shows that the single permutation [s & t] of $S^T P_n\binom{s}{t}$ has the last element e_n, which locates at the current position (not original position) E_{n-1} of e_{s+1}, and e_{s+1} locates at the last position E_n of e_n.

In the example above, if e_n transposes with another e_{n-3} at the current position E_{s+1}, we get the same results. The last element e_n locates at the current position E_{s+1} of e_{n-3}, and e_{n-3} locates at the last position E_n. But [s & t] is one of the permutations of $S^T P_n\binom{s}{t}$, which equals to $(t-1).P_{n-1}\binom{s}{t-1}[s\,\&\,t]$. If we perform the T transposition between e_n and $(t - 1)$ transposed elements in all single permutations [s & (t – 1)] of $P_{n-1}\binom{s}{t-1}$, we get the whole single permutation [s & t] of $S^T P_n \binom{s}{t}$ with the same results. Therefore, we deduce the subpartial permutations $S^T P_n \binom{s}{t}$ has two properties:

- In any single permutation of $S^T P_n \binom{s}{t}$, there is always a transposed element that locates at the last position E_n.
- In any single permutation of $S^T P_n \binom{s}{t}$, the last element e_n always locates at the position of which the order is different from that of the transposed element, which is locating at the last position E_n.

4.3: The Properties of $S^J P_n \binom{s}{t}$

Based on the third rule, the single permutation [s & t] of $S^J P_n \binom{s}{t}$ is created by a stationed element e_n that joins the single permutation [s – 1 & t] of $P_{n-1}\binom{s-1}{t}$ without transposition. So the third rule adds e_n to [s – 1 & t] to get [s & t], and e_n always stations at its original position E_n. Let's take an example that e_n joins a special single permutation of $P_{n-1}\binom{s-1}{t}$, as Diagram (3) describes.

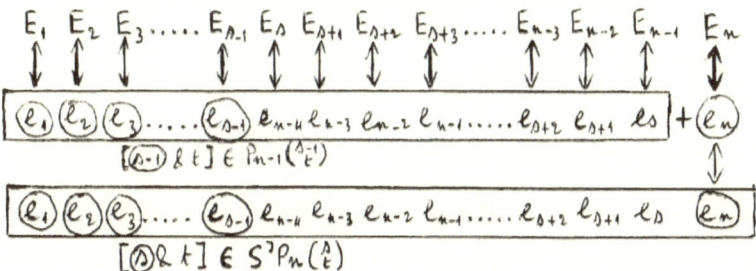

Diagram (3) shows that the single permutation [s & t] of $S^J P_n \binom{s}{t}$ has the last element e_n, which always locates at the last position E_n. But [s & t] is one of the permutations of $S^J P_n \binom{s}{t}$ that equals to $P_{n-1}\binom{s-1}{t}[s \& t]$. If we perform the J rule by adding e_n to all single permutations [s - 1 & t] of $P_{n-1}\binom{s-1}{t}$, we get the whole single permutation [s & t] of $S^J P_n \binom{s}{t}$ with the same results. Therefore, we deduce that the subpartial permutations $S^J P_n \binom{s}{t}$ has one property following:

- In any single permutation of $S^J P_n \binom{s}{t}$, the last element e_n always locates at the last position E_n.

4.4: The Distinction

We first keep in mind that all the single permutation [s & t] of $S^S P_n \binom{s}{t}$, $S^T P_n \binom{s}{t}$, and $S^J P_n \binom{s}{t}$ are created by e_n, which joins to permute with the single permutations [s + 1 & (t - 2)] of $P_{n-1}\binom{s+1}{t-2}$, [s & (t - 1)] of $P_{n-1}\binom{s}{t-1}$, and [s - 1 & t] of $P_{n-1}\binom{s-1}{t}$

, respectively. In the previous subjects, we have selected the special single permutation [s + 1 & (t − 2)], [s & (t − 1)], and [s − 1 & t] that contain the most elements to be coincident of one another because this is the most potential case for the special single permutation [s & t] of those subpartial permutations to be coincident of one another. Otherwise, there are no coincidence between them.

To know whether these single permutations [s & t] of $S^S P_n \binom{s}{t}$, $S^T P_n \binom{s}{t}$, and $S^J P_n \binom{s}{t}$ to be coincident or not, let's take:

- the special single permutation [s & t] of $S^S P_n \binom{s}{t}$ in Diagram (1)
- $S^T P_n \binom{s}{t}$ in Diagram (2)
- $S^J P_n \binom{s}{t}$ in Diagram (3)

Then we will put them together in such a manner that their elements stay at the positions from E_1 to E_n, as Diagram (4) describes.

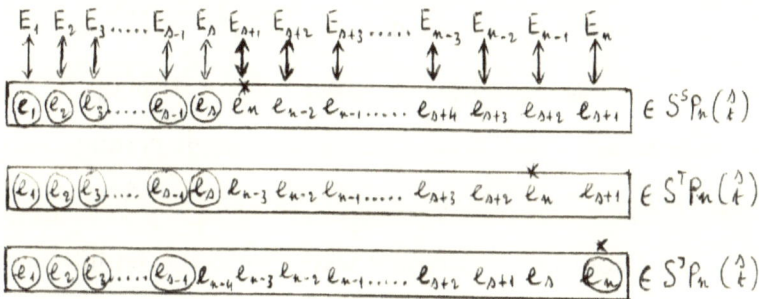

Diagram (4) shows that the last element e_n of three single permutations above locate at different positions. Now we disregard the other elements of [s & t] that coincide one another or not. We only focus on the last element e_n in [s & t] of those subpartial permutations.

- In the single permutation $[s \& t] \in S^S P_n \binom{s}{t}$, element e_n locates at the position E_{s+1}, while its two correspondent elements, one in $[s \& t] \in S^T P_n \binom{s}{t}$ and one in $[s \& t] \in S^J P_n \binom{s}{t}$, get the order ≠ the order of e_n.
- In the single permutation $[s \& t] \in S^T P_n \binom{s}{t}$, element e_n locates at position E_{n-1}, while its two correspondent elements, one in $[s \& t] \in S^S P_n \binom{s}{t}$ and one in $[s \& t] \in S^J P_n \binom{s}{t}$, get the order ≠ the order of e_n.
- In the single permutation $[s \& t] \in S^J P_n \binom{s}{t}$, element e_n locates at its original position E_n, while its two

correspondent elements, one in $\left[s\,\&\,t\right] \in S^S P_n \binom{s}{t}$ and one in $\left[s\,\&\,t\right] \in S^T P_n \binom{s}{t}$, get the order ≠ the order of e_n.

- The results above prove that all three single permutations $\left[s\,\&\,t\right] \in S^S P_n \binom{s}{t}$, $\left[s\,\&\,t\right] \in S^T P_n \binom{s}{t}$, and $\left[s\,\&\,t\right] \in S^J P_n \binom{s}{t}$ do not coincide one another according to the definition of coincidence between two single permutations.

Conclusion

Because we have the following:

- $S^S P_n \binom{s}{t} = (s+1).P_{n-1} \binom{s+1}{t-2} \left[s\,\&\,t\right]$
- $S^T P_n \binom{s}{t} = (t-1).P_{n-1} \binom{s}{t-1} \left[s\,\&\,t\right]$
- $S^J P_n \binom{s}{t} = P_{n-1} \binom{s-1}{t} \left[s\,\&\,t\right]$

That means $S^S P_n \binom{s}{t}$, $S^T P_n \binom{s}{t}$, and $S^J P_n \binom{s}{t}$ consist of numbers of their own single permutations that do not coincide with one another according to the results above. Thus, $S^S P_n \binom{s}{t}$, $S^T P_n \binom{s}{t}$, and $S^J P_n \binom{s}{t}$ are distinctive of each other.

Chapter 5

Relationship between $P_n\binom{s}{t}$ and Its Subpartial Permutations

Preliminary

Hitherto, we have mentioned only the creations and properties of the subpartial permutations $S^S P_n\binom{s}{t}$, $S^T P_n\binom{s}{t}$, and $S^J P_n\binom{s}{t}$ in separating from the partial permutations $P_n\binom{s}{t}$. Therefore, a question arises to us: Is there any relationship between $P_n\binom{s}{t}$ and those subpartial permutations?

5.I: Theorem 2

Consider three species of subpartial permutations as following:

- $S^S P_n\binom{s}{t} = (s+1).P_{n-1}\binom{s+1}{t-2}$ (first rule)
- $S^T P_n\binom{s}{t} = (t-1).P_{n-1}\binom{s}{t-1}$ (second rule)
- $S^J P_n\binom{s}{t} = P_{n-1}\binom{s-1}{t}$ (third rule)

The theorem assumes that the partial permutations $P_n\binom{s}{t}$ equal the sum of those subpartial permutations:

$$P_n\binom{s}{t} = S^S P_n\binom{s}{t} + S^T P_n\binom{s}{t} + S^J P_n\binom{s}{t}$$

Demonstration

The theorem assumption is based on two factual reasons:

- The sum of those subpartial permutations reaches the maximum amount of permutations [s & t] because there are no coincidences between [s & t] of those subpartial permutations according to the properties of subpartial permutations.
- With a careful examination, the author has found no other rules to create some other subpartial permutations that differ from $S^S P_n\binom{s}{t}$, $S^T P_n\binom{s}{t}$, and $S^J P_n\binom{s}{t}$.

However, maybe someone refutes this assumption and asserts there might exist some hidden rules that would create some hidden subpartial permutations. That means:

$$P_n\binom{s}{t} > S^S P_n\binom{s}{t} + S^T P_n\binom{s}{t} + S^J P_n\binom{s}{t}$$

Therefore, we must deal with two contradictory assumptions as:

- $P_n\binom{s}{t} = S^S P_n\binom{s}{t} + S^T P_n\binom{s}{t} + S^J P_n\binom{s}{t}$ (author assumption)
- $P_n\binom{s}{t} > S^S P_n\binom{s}{t} + S^T P_n\binom{s}{t} + S^J P_n\binom{s}{t}$ (refuter assumption)

To know which one is true or false, we have to investigate both of them in the following steps:

1. Just by saying hidden rules or hidden subpartial permutations, the refuter obviously does not know what and where they are. Moreover, he cannot specify them. Eventually, his assumption is baseless and has no proof. Therefore, the author assumption is still dependable because it is based on two factual reasons that we have mentioned above.

2. This is the verification step. To verify those assumptions, we must calculate $P_n\binom{s}{t}$ and the sum $S^S P_n\binom{s}{t} + S^T P_n\binom{s}{t} + S^J P_n\binom{s}{t}$ in relation with n, s, and t: $P_n\binom{s}{t} = {}^nC_s \cdot P_t\binom{o}{t} = \dfrac{n!}{(n-s)!s!} \cdot P_t\binom{o}{t}$ (Theorem 1). The sum of three subpartial permutations is:

$$S^S P_n \binom{s}{t} + S^T P_n \binom{s}{t} + S^J P_n \binom{s}{t}$$
$$(s+1).P_{n-1}\binom{s+1}{t-2} + (t-1).P_{n-1}\binom{s}{t-1} + P_{n-1}\binom{s-1}{t}$$
$$\downarrow \qquad\qquad \downarrow \qquad\qquad \downarrow$$
$$(n \geq 2, t \geq 2) \quad (n \geq 3, t \geq 3) \quad (n \geq 1, s \geq 1, t \neq 1)$$

The calculations start from n = 1 and then n = 2, 3, 4 ... until to any large number as we please.

- **n = 1.** There is one case: $P_1\binom{1}{0} \to s = 1$ and t = 0 (because t $\neq$ 1). $P_n\binom{s}{t} = P_1\binom{1}{0} = 1$ (see the expansions of $P_n\binom{s}{t}$). The sum of three subpartial permutations is as follows.

$$\begin{cases} S^S P_1\binom{1}{0} = (1+1).P_{1-1}\binom{1+1}{0-2} = 2.P_0\binom{2}{-2} = 0 & (\text{no existence}) \\ + & + & + & + \\ S^T P_1\binom{1}{0} = (0-1).P_{1-1}\binom{1}{0-1} = (-1).P_0\binom{1}{-1} = 0 & (\text{no existence}) \\ + & + & + & + \\ S^J P_1\binom{1}{0} = P_{1-1}\binom{1-1}{0} = P_0\binom{0}{0} = 1 & (\text{Theorem 1 corollary}) \\ \text{Sum} & = & 1 \end{cases}$$

Therefore, we have:
$$P_1\binom{1}{0} = S^S P_1\binom{1}{0} + S^T P_1\binom{1}{0} + S^J P_1\binom{1}{0} = 1$$

The result proves that the author assumption (a) is accepted and that of refuter is rejected.

- **n = 2.** There are two cases: $P_2\binom{2}{0}$ and $P_2\binom{0}{2}$:

- **First Case: n = 2, s = 2, and t = 0.** $P_n\binom{s}{t}=P_2\binom{2}{0}=1$ (see the expansions of $P_n\binom{s}{t}$). The sum of three subpartial permutations is as follows.

$$\begin{cases} S^S P_2\binom{2}{0}=(2+1).P_{2-1}\binom{2+1}{0-2})=3.P_1\binom{3}{-2})=0 & (\text{No existence}) \\ + & + & + & + \\ S^T P_2\binom{2}{0}=(0-1).P_{2-1}\binom{2}{0-1})=(-1).P_1\binom{2}{-1})=0 & (\text{No existence}) \\ + & + & + & + \\ S^J P_2\binom{2}{0}= & P_{2-1}\binom{2-1}{0})= & P_1\binom{1}{0})=1 & (\text{See expansions of } P_n\binom{s}{t})) \\ \text{Sum} & = & 1 \end{cases}$$

Therefore, we have:

$$P_2\binom{2}{0}=S^S P_2\binom{2}{0}+S^T P_2\binom{2}{0}+S^J P_2\binom{2}{0}=1$$

So the assumption (a) is accepted, and (b) is rejected.

- **Second Case: n = 2, s = 0, and t = 2.** $P_n\binom{s}{t}=P_2\binom{0}{2}=1$ (see the expansions of $P_n\binom{s}{t}$). The sum of three subpartial permutations is as follows.

$$\begin{cases} S^S P_2\binom{0}{2}=(0+1).P_{2-1}\binom{0+1}{2-2})=1.P_1\binom{1}{0})=1 & (\text{see expansions of } P_n\binom{s}{t})) \\ + & + & + & + \\ S^T P_2\binom{0}{2}=(2-1).P_{2-1}\binom{0}{2-1})=1.P_1\binom{0}{1})=0 & (\text{No existence}) \\ + & + & + & + \\ S^J P_2\binom{0}{2}= & P_{2-1}\binom{0-1}{2})= & P_1\binom{-1}{2})=0 & (\text{No existence}) \\ \text{Sum} & = & 1 \end{cases}$$

Therefore, we have:

$$P_2\binom{0}{2}=S^S P_2\binom{0}{2}+S^T P_2\binom{0}{2}+S^J P_2\binom{0}{2}=1$$

So the assumption (a) is accepted, and the (b) is rejected.

- $n = 3$. There are three cases: $P_3\binom{3}{0}$, $P_3\binom{1}{2}$, and $P_3\binom{0}{3}$:
  - **First Case: n = 3, s = 3, and t = 0.** $P_n\binom{s}{t} = P_3\binom{3}{0} = 1$ (expansions of $P_n\binom{s}{t}$). The sum is as follows.

$$\begin{cases} S^S P_3\binom{3}{0} = (3+1) \cdot P_{3-1}\binom{3+1}{0-1} = 4 \cdot P_2\binom{4}{-1} = 0 & (\text{No existence}) \\ + & + & + & + \\ S^T P_3\binom{3}{0} = (0-1) \cdot P_{3-1}\binom{3}{0-1} = (-1) \cdot P_2\binom{3}{-1} = 0 & (\text{No existence}) \\ + & + & + & + \\ \underline{S^J P_3\binom{3}{0}} = \quad P_{3-1}\binom{3-1}{0} = \quad P_2\binom{2}{0} = \underline{1} & (\text{Expansions of } P_n\binom{s}{t}) \\ \text{Sum} & = & & 1 \end{cases}$$

Therefore, we have:

$$P_3\binom{3}{0} = S^S P_3\binom{3}{0} + S^T P_3\binom{3}{0} + S^J P_3\binom{3}{0} = 1$$

So the assumption (a) is accepted, and the (b) is rejected.

- **Second Case: n = 3, s = 1, and t = 2.** The partial permutations $P_n\binom{s}{t}$ becomes:

$$P_3\binom{1}{2} = {}^3C_1 \cdot P_2\binom{0}{2} = 3 \cdot 1 = 3 \text{ (Theorem 1)}$$

The sum of three subpartial permutations is as follows.

$$\begin{bmatrix} S^S P_3\binom{1}{2} = (1+1) \cdot P_{3-1}\binom{1+1}{2-2} = 2 \cdot P_2\binom{2}{0} = 2 & \text{(expansions of } P_n\binom{s}{t})) \\ + & + & + & + \\ S^T P_3\binom{1}{2} = (2-1) \cdot P_{3-1}\binom{1}{2-1} = 1 \cdot P_2\binom{1}{1} = 0 & \text{(No existence)} \\ + & + & + & + \\ \underline{S^J P_3\binom{1}{2}} = & P_{3-1}\binom{1-1}{2} = & P_2\binom{0}{2} = \underline{1} & \text{(expansions of } P_n\binom{s}{t})) \\ \text{Sum} & = & & 3 \end{bmatrix}$$

Therefore:

$$P_3\binom{1}{2} = S^S P_3\binom{1}{2} + S^T P_3\binom{1}{2} + S^J P_3\binom{1}{2} = 3$$

The result accepts the assumption (a) and rejects the (b).

- **Third Case: n = 3, s = 0, and t = 3.** $P_n\binom{s}{t} = P_3\binom{0}{3} = 2$ (see Theorem 1 application). The sum of three subpartial permutations is as follows.

$$\begin{bmatrix} S^S P_3\binom{0}{3} = (0+1) \cdot P_{3-1}\binom{0+1}{3-2} = 1 \cdot P_2\binom{1}{1} = 0 & \text{(No existence)} \\ + & + & + & + \\ S^T P_3\binom{0}{3} = (3-1) \cdot P_{3-1}\binom{0}{3-1} = 2 \cdot P_2\binom{0}{2} = 2 \\ + & + & + & + \\ \underline{S^J P_3\binom{0}{3}} = & P_{3-1}\binom{0-1}{3} = & P_2\binom{-1}{3} = \underline{0} & \text{(No existence)} \\ \text{Sum} & = & & 2 \end{bmatrix}$$

Therefore, we get:

$$P_3\binom{0}{3} = S^S P_3\binom{0}{3} + S^T P_3\binom{0}{3} + S^J P_3\binom{0}{3} = 2$$

So the assumption (a) is accepted, and the (b) is rejected.

- **n = 4.** We get four cases:

$P_4\binom{4}{0}$, $P_4\binom{2}{2}$, $P_4\binom{1}{3}$, and $P_4\binom{0}{4}$:

- **First Case: n = 4, s = 4, and t = 0.** $P_n\binom{s}{t} = P_4\binom{4}{0} = 1$ (see the expansions of $P_n\binom{s}{t}$). The sum of three subpartial permutations is as follows.

$$\begin{bmatrix} S^S P_4\binom{4}{0} = (4+1) \cdot P_{4-1}\binom{4+1}{0-1} = 5 \cdot P_3\binom{5}{-1} = 0 & \text{(No existence)} \\ + & + & + & + \\ S^T P_4\binom{4}{0} = (0-1) \cdot P_{4-1}\binom{4}{0-1} = (-1) \cdot P_3\binom{4}{-1} = 0 & \text{(No existence)} \\ + & + & + & + \\ \underline{S^J P_4\binom{4}{0}} = P_{4-1}\binom{4-1}{0} = P_3\binom{3}{0} = \underline{1} & \text{(expansions of } P_n\binom{s}{t}\text{)} \\ \downarrow & & & \downarrow \\ \text{Sum} & = & & 1 \end{bmatrix}$$

Therefore, we get:

$$P_4\binom{4}{0} = S^S P_4\binom{4}{0} + S^T P_4\binom{4}{0} + S^J P_4\binom{4}{0} = 1$$

The result accepts the assumption (a) and rejects the (b).

- **Second Case: n = 4, s = 2, and t = 2.**

$$P_n\binom{s}{t} = P_4\binom{2}{2} = {}^4C_2 \cdot P_2\binom{0}{2} = 6 \text{ (Theorem 1).}$$

The sum of three subpartial permutations is as follows.

$$\begin{bmatrix} S^S P_4\binom{2}{2} = (2+1) \cdot P_{4-1}\binom{2+1}{2-2} = 3 \cdot P_3\binom{3}{0} = 3 & = 3 \\ + & + & + & + & + \\ S^T P_4\binom{2}{2} = (2-1) \cdot P_{4-1}\binom{2}{2-1} = 1 \cdot P_3\binom{2}{1} = 0 & = 0 & \text{(No existence)} \\ + & + & + & + & + \\ \underline{S^J P_4\binom{2}{2}} = P_{4-1}\binom{2-1}{2} = P_3\binom{1}{2} = {}^3C_1 \cdot P_2\binom{0}{2} = \underline{3} & \text{(Theorem 1)} \\ \downarrow & & & & \downarrow \\ \text{Sum} & = & & & 6 \end{bmatrix}$$

Therefore, we have:

$$P_4\binom{2}{2} = S^S P_4\binom{2}{2} + S^T P_4\binom{2}{2} + S^J P_4\binom{2}{2} = 6$$

The result accepts the assumption (a) and rejects the (b).

- **Third Case: n = 4, s = 1, and t = 3.**

$$P_n\binom{s}{t} = P_4\binom{1}{3} = {}^4C_1 \cdot P_3\binom{0}{3} = 4.2 = 8 \text{ (Theorem 1)}.$$

The sum of three subpartial permutations is as follows.

$$\begin{bmatrix} S^S P_4\binom{1}{3} = (1+1) \cdot P_{4-1}\binom{1+1}{3-2} = 2 \cdot P_3\binom{2}{1} = 0 & = 0 \text{ (No existence)} \\ + & + & + & + & + \\ S^T P_4\binom{1}{3} = (3-1) \cdot P_{4-1}\binom{1}{3-1} = 2 \cdot P_3\binom{1}{2} = 2 \cdot {}^3C_1 \cdot P_2\binom{0}{2} = 6 & \text{(Theorem 1)}. \\ + & + & + & + & + \\ \underline{S^J P_4\binom{1}{3}} = & P_{4-1}\binom{1-1}{3} = & P_3\binom{0}{3} = 2 & = 2 \text{ (Theorems applications)} \\ \downarrow \\ \text{Sum} & = & & \downarrow \\ & & & 8 \end{bmatrix}$$

Therefore, we have:

$$P_4\binom{1}{3} = S^S P_4\binom{1}{3} + S^T P_4\binom{1}{3} + S^J P_4\binom{1}{3} = 8$$

So the assumption (a) of author is accepted and the (b) of the refuter is rejected.

- **Fourth Case: n = 4, s = 0, and t = 4.** $P_n\binom{s}{t} = P_4\binom{0}{4} = 9$ (see Theorem 1 application). The sum of three subpartial permutations is as follows.

$$\begin{bmatrix} S^S P_4\binom{0}{4} = (0+1).P_{4-1}\binom{0+1}{4-1} = 1.P_3\binom{1}{3} = {}^3C_1.P_2\binom{0}{2} = 3 & (Theorem\ 1) \\ + & + & + & + & + \\ S^T P_4\binom{0}{4} = (4-1).P_{4-1}\binom{0}{4-1} = 3.P_3\binom{0}{3} = & 6 & = 6 \\ + & + & + & + & + \\ S^J P_4\binom{0}{4} = & P_{4-1}\binom{0-1}{4} = & P_3\binom{-1}{4} = & 0 & = \underbrace{0}_{9}\ (\text{No existence}) \end{bmatrix}$$
Sum

Therefore, we get:
$$P_4\binom{0}{4} = S^S P_4\binom{0}{4} + S^T P_4\binom{0}{4} + S^J P_4\binom{0}{4} = 9$$

The result accepts the assumption (a) and rejects the (b).

3. Because we cannot verify these two assumptions with n at a very large number, so the refuter may interpolate that the partial permutations $P_n\binom{s}{t}$ might have some extra single permutations that causes:

$$P_n\binom{s}{t} > S^S P_n\binom{s}{t} + S^T P_n\binom{s}{t} + S^J P_n\binom{s}{t}\ \text{(Theorem 1)}$$

Therefore, we must prove that $P_n\binom{s}{t}$ has no extra single permutations at all. If $P_n\binom{s}{t}$ might have some extra single permutations, these single permutations must be specified by s stationed elements and t transposed elements. That means these extra single permutations of $P_n\binom{s}{t}$ having the form [s & t] and s + t = n. Let's set n stationary positions from E_1 to E_n, E_n as the last position, and let's set n elements of [s & t] to stay at these positions with each element at each position. The element identified by the order n is the last element e_n of [s & t], and e_n can be transposed to any positions. In any single permutation [s & t], the s stationed and the t transposed elements are scattered randomly among n positions aforesaid. We already recognize that the location of the last element e_n of [s & t] and the last position E_n determine the properties of three subpartial permutations aforesaid. These subpartial permutations consist of their own single permutations with the same form [s & t] of which the properties are distinctive as the diagram describes.

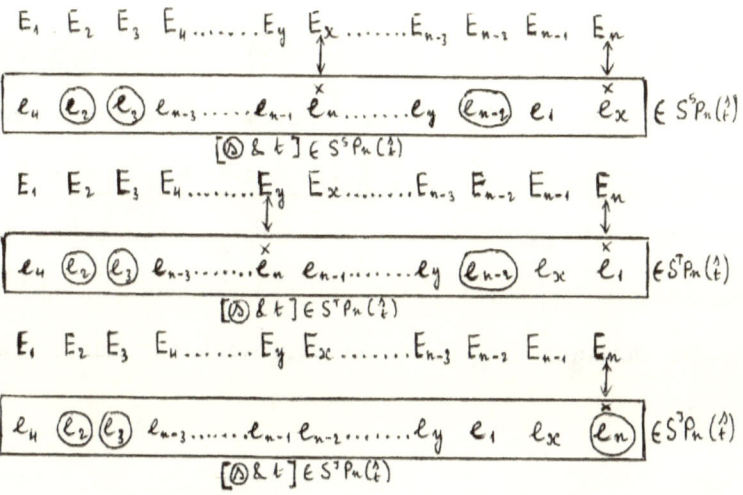

If we analyze any one extra single permutation [s & t], it shows that the last element e_n of [s & t] must locate at one of three following positions. The last element e_n of [s & t] locates at the original position E_x of the transposed element e_x, which is locating at the last position E_n (the original position of e_n). In this case, the extra single permutation [s & t] belongs to the subpartial permutations $S^S P_n \binom{s}{t}$ according to the properties of $S^S P_n \binom{s}{t}$. The last element e_n of [s & t] locates at the position E_y of which the order y is different from the order of the transposed element, which is locating at the last position E_n. In this case, the extra single permutation [s & t] belongs to the subpartial permutations $S^T P_n \binom{s}{t}$ according to the properties of $S^T P_n \binom{s}{t}$. The last element of [s & t] locates at the last position E_n (it original position). In this case, the extra single permutation [s & t] belongs to the subpartial permutations $S^J P_n \binom{s}{t}$ according to the property of $S^J P_n \binom{s}{t}$. The results above prove that any single permutations $[s \& t] \in P_n \binom{s}{t}$ must belong to either $S^S P_n \binom{s}{t}$, $S^T P_n \binom{s}{t}$, or $S^J P_n \binom{s}{t}$. Therefore, the partial permutations $P_n \binom{s}{t}$ have no extra single permutations at all.

Conclusion

Because all three steps above reject the assumption of the refuter in all aspects, the assumption of the theorem is therefore true:
$$P_n\binom{s}{t} = S^S P_n\binom{s}{t} + S^T P_n\binom{s}{t} + S^J P_n\binom{s}{t}$$

Corollary

According to Theorem 2, we have:
$$P_n\binom{s}{t} = S^S P_n\binom{s}{t} + S^T P_n\binom{s}{t} + S^J P_n\binom{s}{t}.$$

But the permutation rules give us:

- $S^S P_n\binom{s}{t} = (s+1).P_{n-1}\binom{s+1}{t-2}$ (first rule)
- $S^T P_n\binom{s}{t} = (t-1).P_{n-1}\binom{s}{t-1}$ (second rule)
- $S^J P_n\binom{s}{t} = P_{n-1}\binom{s-1}{t}$ (third rule)

Therefore, we get:
$$P_n\binom{s}{t} = (s+1).P_{n-1}\binom{s+1}{t-2} + (t-1).P_{n-1}\binom{s}{t-1} + P_{n-1}\binom{s-1}{t}$$
if $s = 0 \to t = n$

The partial permutations $P_n\binom{s}{t}$ becomes:
$$P_n\binom{0}{n} = 1.P_{n-1}\binom{1}{n-2} + (n-1).P_{n-1}\binom{0}{n-1} + P_{n-1}\binom{-1}{n}$$

But:

$$P_{n-1}\binom{1}{n-2} = {}^{n-1}C_1 \cdot P_{n-2}\binom{0}{n-2} \text{ (Theorem 1)}$$

$$= (n-1).P_{n-2}\binom{0}{n-2} \text{ and } P_{n-1}\binom{-1}{n} = 0 \text{ (no existence)}$$

Therefore, we can rewrite:

$$P_n\binom{0}{n} = (n-1).P_{n-2}\binom{0}{n-2} + (n-1).P_{n-1}\binom{0}{n-1} + 0$$

If we simplify and rearrange, we get the relationship:

$$P_n\binom{0}{n} = (n-1).\left[P_{n-1}\binom{0}{n-1} + P_{n-2}\binom{0}{n-2}\right]$$

Remark

This relationship cannot determine the amount of single permutation $[o\,\&\,n] \in P_n\binom{0}{n}$ because we do not know either the amount of single permutation $[o\,\&\,(n-1)] \in P_{n-1}\binom{0}{n-1}$ or the amount of single permutation $[o\,\&\,(n-2)] \in P_{n-2}\binom{0}{n-2}$. However, it's helpful in discovering the general formula of $P_n\binom{0}{n}$.

5.2: Expansion Method

At the previous subject, the author has defined the expansion of the general permutation nPn = n! and that of the partial permutations $P_n \binom{s}{t}$. The author also has expanded some low-grade permutation $P_n \binom{s}{t}$ by experience. Now we present a method to expand nPn and $P_n \binom{s}{t}$ by using the permutation rules in connection with Theorem 2. This method provides a specific and unique way to expand nPn and $P_n \binom{s}{t}$. Let's expand the general permutation nPn = n! and the partial permutation $P_n \binom{s}{t}$ from the first grade to the nth grade as following:

Part A: Diagram I

$$1! \\ \parallel \\ P_1\binom{1}{0} = [e_1] \to e_2 \xrightarrow{J}_{Str.} [e_1 e_2] = P_1\binom{1}{0} = S^J P_2\binom{2}{0} = P_2\binom{2}{0} = 1 \\ + \quad + \quad + \quad + \quad + \\ [e_2\, e_1] = 1.P_1\binom{1}{0} = S^S P_2\binom{0}{2} = P_2\binom{0}{2} = 1 \\ \underbrace{\qquad\qquad}_{2\ permutations}$$

$$2! \\ \parallel \\ \dots \\ 2! = 2$$

Explanation

- Expansion of $1! = P_1\binom{1}{0} = [e_1]$ consists of one single permutation with one stationed element e_1.

- The permutation 2! is created by a stationed element e_2, which joins $P_1\binom{1}{0}$ in two rules:

  - e_2, which joins $P_1\binom{1}{0}$ with no transpositions J, produces:
    $$[e_1 e_2] = S^J P_2\binom{2}{0} = P_2\binom{2}{0}$$

  - e_2, which joins $P_1\binom{1}{0}$ by 5tr. with e_1, produces

$$[e_2e_1] = S^S P_2 \begin{pmatrix} 0 \\ 2 \end{pmatrix} = P_2 \begin{pmatrix} 0 \\ 2 \end{pmatrix}.$$

Because T = 0, there is no T transposition in this case.

- The expansion of general permutation:

$$2! = P_2 \begin{pmatrix} 2 \\ 0 \end{pmatrix} + P_2 \begin{pmatrix} 0 \\ 2 \end{pmatrix} = [e_1e_2] + [e_2e_1] = 2 = 2!$$

- The expansion of partial permutation:

$$P_2 \begin{pmatrix} 2 \\ 0 \end{pmatrix} = [e_1e_2] = 1 \quad \text{and} \quad P_2 \begin{pmatrix} 0 \\ 2 \end{pmatrix} = [e_2e_1] = 1 = \frac{2!}{2!}$$

Part B: Diagram 2

6 permutations

Explanation

- The general permutation 3! is created by e_3, which joins $2! = P_2 \begin{pmatrix} 2 \\ 0 \end{pmatrix} + P_2 \begin{pmatrix} 0 \\ 2 \end{pmatrix}$ in the following rules:

- e_3, which joins $P_2\binom{2}{0}$ by $\underline{J}$, produces $[e_1e_2e_3] = S^J P_3 \binom{3}{0} = P_3 \binom{3}{0}$. e_3, which joins $P_2\binom{2}{0}$ by 5tr., produces two single permutations:

$$S^S P_3 \binom{1}{2} = \begin{bmatrix} [e_1e_3e_2] \text{ by } e_3 \text{ 5tr. with } e_2 \in P_2\binom{2}{0} \\ [e_3e_2e_1] \text{ by } e_3 \text{ 5tr. with } e_1 \in P_2\binom{2}{0} \end{bmatrix}$$

Because T = 0, thus, there is no Ttr. in this case.

- e_3, which joins $P_2\binom{0}{2}$ by $\underline{J}$, produces $[e_2e_1e_3] = S^J P_3\binom{1}{2}$. e_3, which joins $P_2\binom{0}{2}$ by Ttr., produces two single permutations:

$$S^T P_3 \binom{0}{3} = \begin{bmatrix} [e_2e_3e_1] \text{ by } e_3 \text{ Ttr. with } e_1 \in [e_2e_1] = P_2\binom{0}{2} \\ [e_3e_1e_2] \text{ by } e_3 \text{ Ttr. with } e_2 \in [e_2e_1] = P_2\binom{0}{2} \end{bmatrix}$$

Because S = 0, thus, there is no 5tr. in this case.

- If we summarize (a) and (b) and set all single permutations with the same s stationed and same t transposed elements into their own specific subpartial permutations, we get the following result. The expansions of general permutation 3! and the expansions of partial permutations $P_3\binom{3}{0}$, $P_3\binom{1}{2}$, $P_3\binom{0}{3}$ in the explanation and Diagram 2 are identical.

Part C: Diagram 3

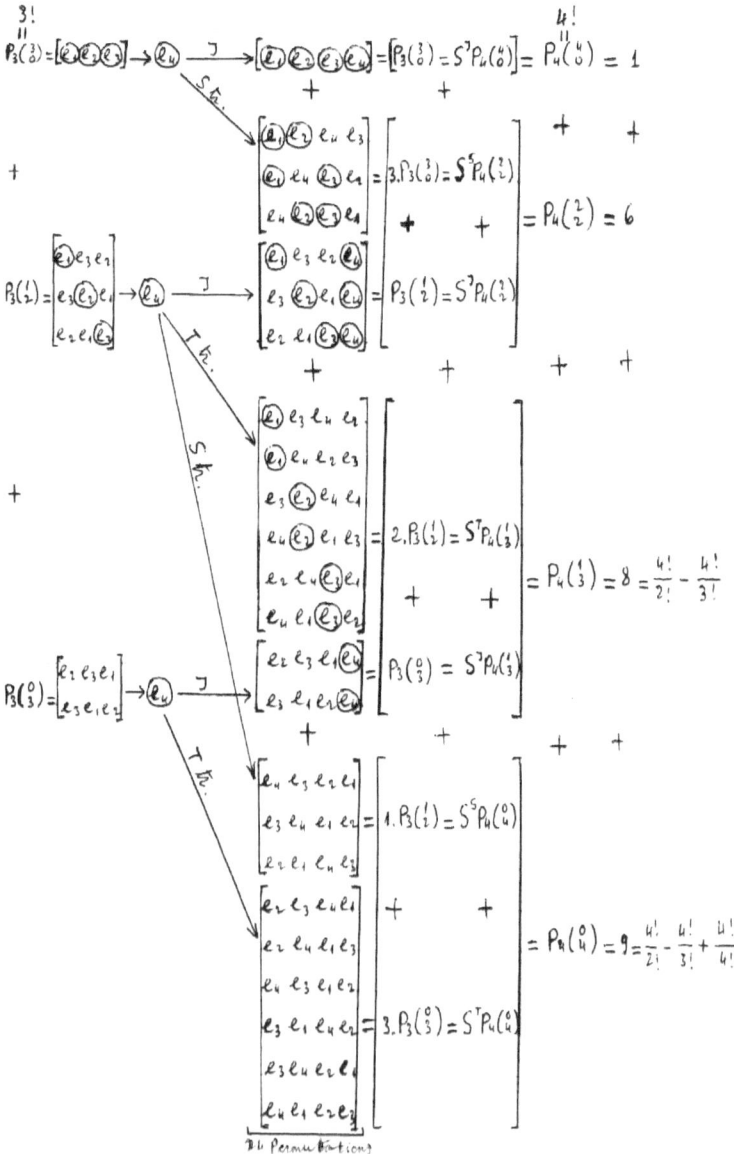

Explanation

- The general permutation 4! is created by e_4, which joins $3! = P_3\binom{3}{0} + P_3\binom{1}{2} + P_3\binom{0}{3}$ in the following rules:
  - e_4, which joins $P_3\binom{3}{0}$ by $\underline{J}$, produces $[e_1e_2e_3e_4] = S^J P_4\binom{4}{0} = P_4\binom{4}{0}$. e_4, which joins $P_3\binom{3}{0}$ by $\underline{5tr.}$ produces three single permutations:

  $$S^S P_4\binom{2}{2} = \begin{bmatrix} [e_1e_2e_4e_3] \text{ by } e_4\underline{5tr.} \text{ with } e_3 \in [e_1e_2e_3] = P_3\binom{3}{0} \\ [e_1e_4e_3e_2] \text{ by } e_4\underline{5tr.} \text{ with } e_2 \in [e_1e_2e_3] = P_3\binom{3}{0} \\ [e_4e_2e_3e_1] \text{ by } e_4\underline{5tr.} \text{ with } e_1 \in [e_1e_2e_3] = P_3\binom{3}{0} \end{bmatrix}.$$

  Because T = 0, thus, there is no $\underline{Ttr.}$ in this case.

  - e_4, which joins $P_3\binom{1}{2}$ by $\underline{J}$, produces three single permutations:

  $$S^J P_4\binom{2}{2} = \begin{bmatrix} [e_1e_3e_2e_4] \text{ by } e_4\underline{J} \text{ with } [e_1e_3e_2] \in P_3\binom{1}{2} \\ [e_3e_2e_1e_4] \text{ by } e_4\underline{J} \text{ with } [e_3e_2e_1] \in P_3\binom{1}{2} \\ [e_2e_1e_3e_4] \text{ by } e_4\underline{J} \text{ with } [e_2e_1e_3] \in P_3\binom{1}{2} \end{bmatrix}$$

  - e_4, which joins $P_3\binom{1}{2}$ by $\underline{Ttr.}$, produces six single permutations:

$$S^T P_4 \begin{pmatrix}1\\3\end{pmatrix} = \begin{bmatrix} [e_1 e_3 e_4 e_2] \text{ by } e_4 \underline{\text{Ttr}}, \text{ with } e_2 \in [e_1 e_3 e_2] \in P_3 \begin{pmatrix}1\\2\end{pmatrix} \\ [e_1 e_4 e_2 e_3] \text{ by } e_4 \underline{\text{Ttr}}, \text{ with } e_3 \in [e_1 e_3 e_2] \in P_3 \begin{pmatrix}1\\2\end{pmatrix} \\ [e_3 e_2 e_4 e_1] \text{ by } e_4 \underline{\text{Ttr}}, \text{ with } e_1 \in [e_3 e_2 e_1] \in P_3 \begin{pmatrix}1\\2\end{pmatrix} \\ [e_4 e_2 e_1 e_3] \text{ by } e_4 \underline{\text{Ttr}}, \text{ with } e_3 \in [e_3 e_2 e_1] \in P_3 \begin{pmatrix}1\\2\end{pmatrix} \\ [e_2 e_4 e_3 e_1] \text{ by } e_4 \underline{\text{Ttr}}, \text{ with } e_1 \in [e_2 e_1 e_3] \in P_3 \begin{pmatrix}1\\2\end{pmatrix} \\ [e_4 e_1 e_3 e_2] \text{ by } e_4 \underline{\text{Ttr}}, \text{ with } e_2 \in [e_2 e_1 e_3] \in P_3 \begin{pmatrix}1\\2\end{pmatrix} \end{bmatrix}$$

- e_4, which joins $P_3 \begin{pmatrix}1\\2\end{pmatrix}$ by $\underline{5\text{tr}}$, produces single permutations:

$$S^S P_4 \begin{pmatrix}0\\4\end{pmatrix} = \begin{bmatrix} [e_4 e_3 e_2 e_1] \text{ by } e_4 \underline{5\text{tr}}, \text{ with } e_1 \in [e_1 e_3 e_2] \in P_3 \begin{pmatrix}1\\2\end{pmatrix} \\ [e_3 e_4 e_1 e_2] \text{ by } e_4 \underline{5\text{tr}}, \text{ with } e_2 \in [e_3 e_2 e_1] \in P_3 \begin{pmatrix}1\\2\end{pmatrix} \\ [e_2 e_1 e_4 e_3] \text{ by } e_4 \underline{5\text{tr}}, \text{ with } e_3 \in [e_2 e_1 e_3] \in P_3 \begin{pmatrix}1\\2\end{pmatrix} \end{bmatrix}$$

- e_4, which joins $P_3 \begin{pmatrix}0\\3\end{pmatrix}$ by $\underline{J}$, produces two single permutations:

$$S^J P_4 \begin{pmatrix}1\\3\end{pmatrix} = \begin{bmatrix} [e_2 e_3 e_1 e_4] \text{ by } e_4 \underline{J} \text{ with } [e_2 e_3 e_1] \in P_3 \begin{pmatrix}0\\3\end{pmatrix} \\ [e_3 e_1 e_2 e_4] \text{ by } e_4 \underline{J} \text{ with } [e_3 e_1 e_2] \in P_3 \begin{pmatrix}0\\3\end{pmatrix} \end{bmatrix}$$

- e_4, which joins $P_3 \begin{pmatrix}0\\3\end{pmatrix}$ by $\underline{\text{Ttr}}$, produces six single permutations:

ANCIENT AND MODERN MATHEMATICS

$$S^T P_4 \begin{pmatrix} 0 \\ 4 \end{pmatrix} = \begin{bmatrix} [e_2 e_3 e_4 e_1] \text{ by } e_4 \underline{\text{Ttr.}} \text{ with } e_1 \in [e_2 e_3 e_1] \in P_3 \begin{pmatrix} 0 \\ 3 \end{pmatrix} \\ [e_2 e_4 e_1 e_3] \text{ by } e_4 \underline{\text{Ttr.}} \text{ with } e_3 \in [e_2 e_3 e_1] \in P_3 \begin{pmatrix} 0 \\ 3 \end{pmatrix} \\ [e_4 e_3 e_1 e_2] \text{ by } e_4 \underline{\text{Ttr.}} \text{ with } e_2 \in [e_2 e_3 e_1] \in P_3 \begin{pmatrix} 0 \\ 3 \end{pmatrix} \\ [e_3 e_1 e_4 e_2] \text{ by } e_4 \underline{\text{Ttr.}} \text{ with } e_2 \in [e_3 e_1 e_2] \in P_3 \begin{pmatrix} 0 \\ 3 \end{pmatrix} \\ [e_3 e_4 e_2 e_1] \text{ by } e_4 \underline{\text{Ttr.}} \text{ with } e_1 \in [e_3 e_1 e_2] \in P_3 \begin{pmatrix} 0 \\ 3 \end{pmatrix} \\ [e_4 e_1 e_2 e_3] \text{ by } e_4 \underline{\text{Ttr.}} \text{ with } e_3 \in [e_3 e_1 e_2] \in P_3 \begin{pmatrix} 0 \\ 3 \end{pmatrix} \end{bmatrix}$$

Because S = 0, thus, there is no $\underline{5\text{tr.}}$ in this case.

- Summarize (a), (b), and (c). If we set all single permutations as having the same s stationed and same t transposed elements into their own specific subpartial permutations, we get the expansions of general permutation 4! and the expansions of partial permutations $P_4 \begin{pmatrix} 4 \\ 0 \end{pmatrix}$, $P_4 \begin{pmatrix} 2 \\ 2 \end{pmatrix}$, $P_4 \begin{pmatrix} 1 \\ 3 \end{pmatrix}$, and $P_4 \begin{pmatrix} 0 \\ 4 \end{pmatrix}$ in the explanation and Diagram 3 are identical.

Part D:

Due to the limited format of paper, we may not expand the general permutations 5! into 120 single permutations. However, we can expand it into partial permutations. After that, we can expand any partial permutations into single permutations if we would like to do so. Like the expansions of the general permutation 4!, the expansions of 5! are created by a stationed element e_5, which joins $4! = P_4\binom{4}{0} + P_4\binom{2}{2} + P_4\binom{1}{3} + P_4\binom{0}{4}$ in the same manner that the previous diagrams has described.

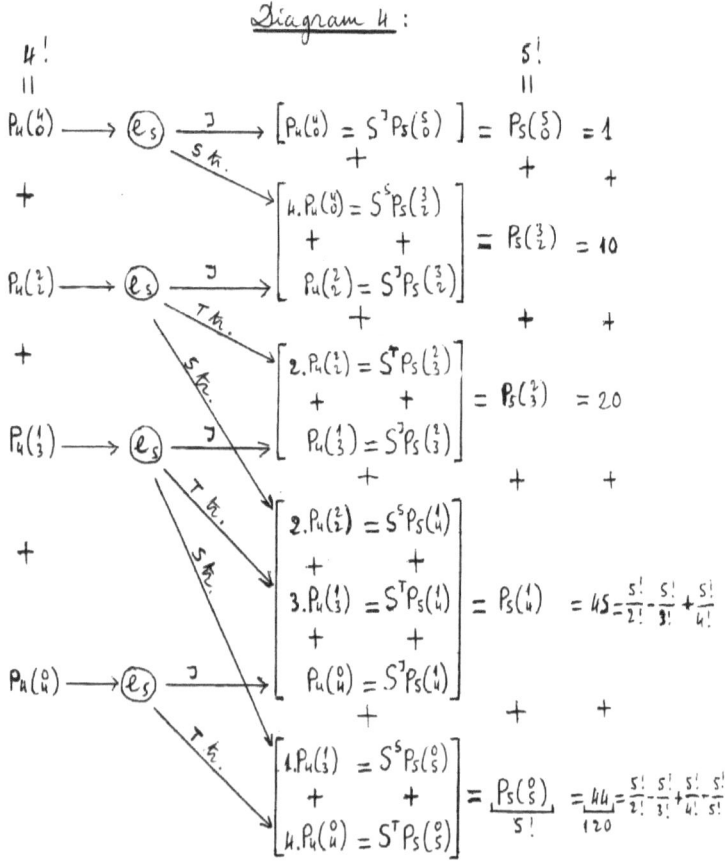

Explanation

The relationship of partial permutations and subpartial permutations in the diagram are based on the permutation rules in connection with Theorem 1 and Theorem 2.

- $P_5\binom{5}{0} = S^J P_5\binom{5}{0} = P_4\binom{4}{0} = 1$ (third rule)
- $P_5\binom{3}{2} = S^S P_5\binom{3}{2} + S^J P_5\binom{3}{2}$ (Theorem 2)
  - $= 4.P_4\binom{4}{0} + P_4\binom{2}{2}$ (first and third rules)
  - $= 4.1 + 6 = 10$ (Theorem 1)
- $P_5\binom{2}{3} = S^T P_5\binom{2}{3} + S^J P_5\binom{2}{3}$ (Theorem 2)
  - $= 2.P_4\binom{2}{2} + P_4\binom{1}{3}$ (second and third rules)
  - $= 2.6 + 8 = 20$ (Theorem 1)
- $P_5\binom{1}{4} = S^S P_5\binom{1}{4} + S^T P_5\binom{1}{4} + S^J P_5\binom{1}{4}$ (Theorem 2)
  - $= 2.P_4\binom{2}{2} + 3.P_4\binom{1}{3} + P_4\binom{0}{4}$ (first, second, and third rules)
  - $= 2.6 + 3.8 + 9 = 45$ (Theorem 1)
- $P_5\binom{0}{5} = S^S P_5\binom{0}{5} + S^T P_5\binom{0}{5}$ (Theorem 2)
  - $= 1.P_4\binom{1}{3} + 4.P_4\binom{0}{4}$ (first and second rules)
  - $= 1.8 + 4.9 = 44$ (Theorem 1)

The expansions of 5! is as follows:

$$5! = P_5\binom{5}{0} + P_5\binom{3}{2} + P_5\binom{2}{3} + P_5\binom{1}{4} + P_5\binom{0}{5}$$
$$= 1 + 10 + 20 + 45 + 44 = 120$$

For the expansions of $P_n\binom{s}{t}$, as an option, let's expand $P_5\binom{1}{4}$ and $P_5\binom{0}{5}$:

A. The partial permutation $P_5\begin{pmatrix}1\\4\end{pmatrix}$ is created as the diagram describes.

$$\begin{array}{c}P_4\begin{pmatrix}2\\2\end{pmatrix}\\+\\P_4\begin{pmatrix}1\\3\end{pmatrix}\\+\\P_4\begin{pmatrix}0\\4\end{pmatrix}\end{array} \begin{array}{c}\longrightarrow \ell_5 \xrightarrow{Sh.}\\ \longrightarrow \ell_5 \xrightarrow{Th.}\\ \longrightarrow \ell_5 \xrightarrow{J}\end{array} \begin{bmatrix}2.P_4\begin{pmatrix}2\\2\end{pmatrix}=S^S P_5\begin{pmatrix}1\\4\end{pmatrix}\\+\\3.P_4\begin{pmatrix}1\\3\end{pmatrix}=S^T P_5\begin{pmatrix}1\\4\end{pmatrix}\\+\\P_4\begin{pmatrix}0\\4\end{pmatrix}=S^J P_5\begin{pmatrix}1\\4\end{pmatrix}\end{bmatrix}=P_5\begin{pmatrix}1\\4\end{pmatrix}$$

Diagram 3 gives us the expansions of $P_4\begin{pmatrix}2\\2\end{pmatrix}$, $P_4\begin{pmatrix}1\\3\end{pmatrix}$, and $P_4\begin{pmatrix}0\\4\end{pmatrix}$. Therefore, we can deduce the expansions of $P_5\begin{pmatrix}1\\4\end{pmatrix}$ as following:

$$P_4\begin{pmatrix}2\\2\end{pmatrix} = \begin{bmatrix}\cdots\end{bmatrix} \xrightarrow{\ell_5} \xrightarrow{S\,h.} \begin{bmatrix}\cdots\end{bmatrix} = S^S P_S\begin{pmatrix}1\\4\end{pmatrix}$$

$$+$$

$$P_4\begin{pmatrix}1\\3\end{pmatrix} = \begin{bmatrix}\cdots\end{bmatrix} \xrightarrow{\ell_5} \xrightarrow{T\,h.} \begin{bmatrix}\cdots\end{bmatrix} = S^T P_S\begin{pmatrix}1\\4\end{pmatrix} = P_S\begin{pmatrix}1\\4\end{pmatrix} = 4S$$

$$+$$

$$P_4\begin{pmatrix}0\\4\end{pmatrix} = \begin{bmatrix}\ell_4\,\ell_3\,\ell_2\,\ell_1\\ \ell_3\,\ell_4\,\ell_1\,\ell_2\\ \ell_2\,\ell_1\,\ell_4\,\ell_3\\ \ell_2\,\ell_3\,\ell_4\,\ell_1\\ \ell_1\,\ell_4\,\ell_1\,\ell_3\\ \ell_4\,\ell_3\,\ell_1\,\ell_1\\ \ell_3\,\ell_1\,\ell_4\,\ell_2\\ \ell_3\,\ell_4\,\ell_2\,\ell_1\\ \ell_4\,\ell_1\,\ell_2\,\ell_3\end{bmatrix} \xrightarrow{\ell_5} \xrightarrow{J} \begin{bmatrix}\cdots\end{bmatrix} = S^J P_S\begin{pmatrix}1\\4\end{pmatrix}$$

B. The $P_5\begin{pmatrix}0\\5\end{pmatrix}$ is created as the diagram describes.

$$\begin{array}{c}P_4\begin{pmatrix}1\\3\end{pmatrix} \longrightarrow \ell_5 \xrightarrow{S\,h.}\\ +\\ P_4\begin{pmatrix}0\\4\end{pmatrix} \longrightarrow \ell_5 \xrightarrow{T\,h.}\end{array} \begin{bmatrix}1.\ P_4\begin{pmatrix}1\\3\end{pmatrix} = S^S P_S\begin{pmatrix}0\\5\end{pmatrix}\\ +\qquad\qquad +\\ 4.\ P_4\begin{pmatrix}0\\4\end{pmatrix} = S^T P_S\begin{pmatrix}0\\5\end{pmatrix}\end{bmatrix} = P_S\begin{pmatrix}0\\5\end{pmatrix}$$

Because we know the expansions of $P_4\binom{1}{3}$ and $P_4\binom{0}{4}$, we thus deduce the expansions of $P_5\binom{0}{5}$ as following:

$$P_4\binom{1}{3} = \begin{bmatrix} \boxed{\ell_1}\ \ell_3\ \ell_4\ \ell_2 \\ \boxed{\ell_1}\ \ell_4\ \ell_2\ \ell_3 \\ \ell_3\ \boxed{\ell_2}\ \ell_4\ \ell_1 \\ \ell_4\ \boxed{\ell_2}\ \ell_1\ \ell_3 \\ \ell_2\ \ell_4\ \boxed{\ell_3}\ \ell_1 \\ \ell_4\ \ell_1\ \boxed{\ell_3}\ \ell_2 \\ \ell_2\ \ell_3\ \ell_1\ \boxed{\ell_4} \\ \ell_3\ \ell_1\ \ell_2\ \boxed{\ell_4} \end{bmatrix} \longrightarrow \boxed{\ell_5} \xrightarrow{S\,th.} \begin{bmatrix} \ell_5\ \ell_3\ \ell_4\ \ell_2\ \ell_1 \\ \ell_5\ \ell_4\ \ell_2\ \ell_3\ \ell_1 \\ \ell_3\ \ell_5\ \ell_4\ \ell_1\ \ell_2 \\ \ell_4\ \ell_5\ \ell_1\ \ell_3\ \ell_2 \\ \ell_2\ \ell_4\ \ell_5\ \ell_1\ \ell_3 \\ \ell_4\ \ell_1\ \ell_5\ \ell_2\ \ell_3 \\ \ell_2\ \ell_3\ \ell_1\ \ell_5\ \ell_4 \\ \ell_3\ \ell_1\ \ell_2\ \ell_5\ \ell_4 \end{bmatrix} = S^S P_5\binom{0}{5}$$

$$P_4\binom{0}{4} = \begin{bmatrix} \ell_4\ \ell_3\ \ell_2\ \ell_1 \\ \ell_3\ \ell_4\ \ell_1\ \ell_2 \\ \ell_2\ \ell_1\ \ell_4\ \ell_3 \\ \ell_2\ \ell_1\ \ell_3\ \ell_4 \\ \ell_1\ \ell_4\ \ell_2\ \ell_3 \\ \ell_4\ \ell_3\ \ell_1\ \ell_2 \\ \ell_3\ \ell_1\ \ell_4\ \ell_2 \\ \ell_3\ \ell_4\ \ell_2\ \ell_1 \\ \ell_4\ \ell_1\ \ell_2\ \ell_3 \end{bmatrix} \longrightarrow \boxed{\ell_5} \xrightarrow{T\,th.} \begin{bmatrix} \ell_4\ \ell_3\ \ell_2\ \ell_5\ \ell_1 \\ \ell_4\ \ell_3\ \ell_5\ \ell_1\ \ell_2 \\ \ell_4\ \ell_5\ \ell_2\ \ell_1\ \ell_3 \\ \ell_5\ \ell_3\ \ell_2\ \ell_1\ \ell_4 \\ \ell_3\ \ell_4\ \ell_2\ \ell_5\ \ell_2 \\ \ell_3\ \ell_4\ \ell_5\ \ell_2\ \ell_1 \\ \ell_3\ \ell_5\ \ell_1\ \ell_2\ \ell_4 \\ \ell_5\ \ell_4\ \ell_1\ \ell_2\ \ell_3 \\ \ell_2\ \ell_1\ \ell_4\ \ell_5\ \ell_3 \\ \ell_2\ \ell_1\ \ell_5\ \ell_3\ \ell_4 \\ \ell_2\ \ell_5\ \ell_4\ \ell_3\ \ell_1 \\ \ell_5\ \ell_1\ \ell_4\ \ell_3\ \ell_2 \\ \ell_2\ \ell_3\ \ell_4\ \ell_5\ \ell_1 \\ \ell_2\ \ell_3\ \ell_5\ \ell_1\ \ell_4 \\ \ell_2\ \ell_5\ \ell_1\ \ell_4\ \ell_3 \\ \ell_5\ \ell_3\ \ell_4\ \ell_1\ \ell_2 \\ \ell_1\ \ell_4\ \ell_5\ \ell_3\ \ell_2 \\ \ell_5\ \ell_4\ \ell_1\ \ell_3\ \ell_2 \\ \ell_1\ \ell_4\ \ell_5\ \ell_3\ \ell_2 \\ \ell_4\ \ell_5\ \ell_1\ \ell_2\ \ell_3 \\ \ell_5\ \ell_3\ \ell_1\ \ell_2\ \ell_4 \\ \ell_3\ \ell_1\ \ell_5\ \ell_2\ \ell_4 \\ \ell_3\ \ell_5\ \ell_2\ \ell_4 \\ \ell_5\ \ell_3\ \ell_4\ \ell_2\ \ell_1 \\ \ell_3\ \ell_4\ \ell_2\ \ell_5\ \ell_1 \\ \ell_3\ \ell_4\ \ell_2\ \ell_1\ \ell_5 \\ \ell_5\ \ell_4\ \ell_1\ \ell_3\ \ell_4 \end{bmatrix} = S^T P_5\binom{0}{5}$$

$$+\ \Big] = P_5\binom{0}{5} = 44$$

Chapter 6

Solution Of $P_n \binom{o}{n}$

Preliminary

As we are reminded, the partial permutations $P_n \binom{o}{n}$ is the master key to solve any partial permutations $P_n \binom{s}{t}$. So we must find out any ways to solve $P_n \binom{o}{n}$.

6.1: General Formula

A. Let's calculate the difference of two numbers below:

$\left[P_{n_i} \binom{o}{n_i} - n_i . P_{n_{i-1}} \binom{o}{n_{i-1}} \right]$ (briefly named the difference), where $n_i = n, n-1, n-2 \ldots 5, 4, 3, 2$.

- $n_i = n$. The corollary of Theorem 2 gives us:

$$P_n \binom{o}{n} = (n-1) . \left[P_{n-1} \binom{o}{n-1} + P_{n-2} \binom{o}{n-2} \right]$$

If we add $P_{n-1} \binom{o}{n-1}$ to both members of this relationship, we get:

$$P_{n-1} \binom{o}{n-1} + P_n \binom{o}{n} = P_{n-1} \binom{o}{n-1} + (n-1) . P_{n-1} \binom{o}{n-1} + (n-1) . P_{n-2} \binom{o}{n-2}$$

or

$$P_{n-1}\binom{o}{n-1}+P_n\binom{o}{n}=n.P_{n-1}\binom{o}{n-1}+(n-1).P_{n-2}\binom{o}{n-2}$$

If we move $P_{n-1}\binom{o}{n-1}$ to the second member and $n.P_{n-1}\binom{o}{n-1}$ to the first member of the relationship, we get:

$$\left[P_n\binom{o}{n}-n.P_{n-1}\binom{o}{n-1}\right]=-\left[P_{n-1}\binom{o}{n-1}-(n-1).P_{n-2}\binom{o}{n-2}\right]$$

- $n_i = n - 1$. The corollary of Theorem 2 also gives us:

$$P_{n-1}\binom{o}{n-1}=(n-2).\left[P_{n-2}\binom{o}{n-2}+P_{n-3}\binom{o}{n-3}\right]$$

If we add $P_{n-2}\binom{o}{n-2}$ to both members of this relationship as below, we get:

$$P_{n-2}\binom{o}{n-2}+P_{n-1}\binom{o}{n-1}=P_{n-2}\binom{o}{n-2}+(n-2).P_{n-2}\binom{o}{n-2}+(n-2).P_{n-3}\binom{o}{n-3}$$

or

$$P_{n-2}\binom{o}{n-2}+P_{n-1}\binom{o}{n-1}=(n-1).P_{n-2}\binom{o}{n-2}+(n-2).P_{n-3}\binom{o}{n-3}$$

If we move $P_{n-1}\binom{o}{n-1}$ to the second member and $(n-2).P_{n-3}\binom{o}{n-3}$ to the first member of the relationship, we get the result:

$$\left[P_{n-2}\binom{o}{n-2}-(n-2).P_{n-3}\binom{o}{n-3}\right]=-\left[P_{n-1}\binom{o}{n-1}-(n-1).P_{n-2}\binom{o}{n-2}\right]$$

- $n_i = n - 2$. We also have:

$$P_{n-2}\binom{o}{n-2}=(n-3).\left[P_{n-3}\binom{o}{n-3}+P_{n-4}\binom{o}{n-4}\right]$$

If we add $P_{n-3}\binom{o}{n-3}$ to both members of this relationship, we get:

$$P_{n-3}\binom{o}{n-3}+P_{n-2}\binom{o}{n-2}=P_{n-3}\binom{o}{n-3}+(n-3).P_{n-3}\binom{o}{n-3}+(n-3).P_{n-4}\binom{o}{n-4}$$

or

$$P_{n-3}\binom{o}{n-3}+P_{n-2}\binom{o}{n-2}=(n-2).P_{n-3}\binom{o}{n-3}+(n-3).P_{n-4}\binom{o}{n-4}$$

If we move $P_{n-3}\binom{o}{n-3}$ to the second member and $(n-2).P_{n-3}\binom{o}{n-3}$ to the first member of the relationship, we get the result:

$$\left[P_{n-2}\binom{o}{n-2}-(n-2).P_{n-3}\binom{o}{n-3}\right]=-\left[P_{n-3}\binom{o}{n-3}-(n-3).P_{n-4}\binom{o}{n-4}\right]$$

- $n_i = n - 3$. We also get:

$$P_{n-3}\binom{o}{n-3}=(n-4).\left[P_{n-4}\binom{o}{n-4}+P_{n-5}\binom{o}{n-5}\right]$$

If we add $P_{n-4}\binom{o}{n-4}$ to both members of this relationship, we get:

$$P_{n-4}\binom{o}{n-4}+P_{n-3}\binom{o}{n-3}=P_{n-4}\binom{o}{n-4}+(n-4).P_{n-4}\binom{o}{n-4}+(n-4).P_{n-5}\binom{o}{n-5}$$

or

$$P_{n-4}\binom{o}{n-4}+P_{n-3}\binom{o}{n-3}=(n-3).P_{n-4}\binom{o}{n-4}+(n-4).P_{n-5}\binom{o}{n-5}$$

If we move $P_{n-3}\binom{o}{n-3}$ to the second member and $(n-4).P_{n-5}\binom{o}{n-5}$ to the first member of the relationship, we get the result:

$$\left[P_{n-4}\binom{0}{n-4}-(n-4).P_{n-5}\binom{0}{n-5}\right]=-\left[P_{n-3}\binom{0}{n-3}-(n-3).P_{n-4}\binom{0}{n-4}\right]$$

- $n_i = n - 4 \ldots 5, 4, 3, 2$. If we use the same method, we get the results from $n_i = n - 4$ down to $n_i = 2$:

* - $n_i = n-4$:
$$\left[P_{n-4}\binom{0}{n-4}-(n-4).P_{n-5}\binom{0}{n-5}\right]=-\left[P_{n-5}\binom{0}{n-5}-(n-5).P_{n-6}\binom{0}{n-6}\right]$$

----

* - $n_i = 5$:
$$\left[P_4\binom{0}{4}-4.P_3\binom{0}{3}\right]=-\left[P_5\binom{0}{5}-5.P_4\binom{0}{4}\right]$$

* - $n_i = 4$:
$$\left[P_4\binom{0}{4}-4.P_3\binom{0}{3}\right]=-\left[P_3\binom{0}{3}-3.P_2\binom{0}{2}\right]$$

* - $n_i = 3$:
$$\left[P_2\binom{0}{2}-2.P_1\binom{0}{1}\right]=-\left[P_3\binom{0}{3}-3.P_2\binom{0}{2}\right]$$

* - $n_i = 2$:
$$\left[P_2\binom{0}{2}-2.P_1\binom{0}{1}\right]=-\left[P_1\binom{0}{1}-1.P_0\binom{0}{0}\right]$$

B. If we summarize all the results from $n_i = n$ down to $n_i = 2$ and arrange the equalities of those differences in vertical column, we get the table 1 as following.

$$+\left[P_n\binom{0}{n} - n \cdot P_{n-1}\binom{0}{n-1}\right]$$
$$\|$$
$$-\left[P_{n-1}\binom{0}{n-1} - (n-1) \cdot P_{n-2}\binom{0}{n-2}\right]$$
$$\|$$
$$+\left[P_{n-2}\binom{0}{n-2} - (n-2) \cdot P_{n-3}\binom{0}{n-3}\right]$$
$$\|$$
$$-\left[P_{n-3}\binom{0}{n-3} - (n-3) \cdot P_{n-4}\binom{0}{n-4}\right]$$
$$\|$$
$$+\left[P_{n-4}\binom{0}{n-4} - (n-4) \cdot P_{n-5}\binom{0}{n-5}\right]$$
$$\|$$
$$-\left[P_{n-5}\binom{0}{n-5} - (n-5) \cdot P_{n-6}\binom{0}{n-6}\right]$$
$$\|$$
$$-\left[P_5\binom{0}{5} - 5 \cdot P_4\binom{0}{4}\right]$$
$$\|$$
$$+\left[P_4\binom{0}{4} - 4 \cdot P_3\binom{0}{3}\right]$$
$$\|$$
$$-\left[P_3\binom{0}{3} - 3 \cdot P_2\binom{0}{2}\right]$$
$$\|$$
$$+\left[P_2\binom{0}{2} - 2 \cdot P_1\binom{0}{1}\right]$$

The table shows that all those differences have alternate signs from one to another. We also recognize the following:

- All the differences $\left[P_{n_i}\binom{0}{n_i} - n_i \cdot P_{n_{i-1}}\binom{0}{n_{i-1}}\right]$ with the highest-grade partial permutations $P_{n_i}\binom{0}{n_i}$

of which n_i is an even has the plus sign (+), such as $+\left[P_4\binom{0}{4}-4.P_3\binom{0}{3}\right]$ and $+\left[P_2\binom{0}{2}-2.P_1\binom{0}{1}\right]$.

- All the differences with the highest-grade partial permutations $P_{n_i}\binom{0}{n_i}$ of which n_i is an odd has the minus sign (−), such as $-\left[P_5\binom{0}{5}-5.P_4\binom{0}{4}\right]$ and $-\left[P_3\binom{0}{3}-3.P_2\binom{0}{2}\right]$.

On the table 1, all those differences equal to one another. Therefore, each must be a constant K.

We already know:

$$P_2\binom{0}{2}=1,\ P_3\binom{0}{3}=2,\ P_4\binom{0}{4}=9,\ \text{and}\ P_5\binom{0}{5}=44$$

Therefore, we deduce.

$$-\left[P_5\binom{0}{5}-5.P_4\binom{0}{4}\right] = -[44-5.9] = 1$$
$$\|\qquad\qquad\qquad\|$$
$$+\left[P_4\binom{0}{4}-4.P_3\binom{0}{3}\right] = +[9-4.2] = 1$$
$$\|\qquad\qquad\qquad\|$$
$$-\left[P_3\binom{0}{3}-3.P_2\binom{0}{2}\right] = -[2-3.1] = 1$$
$$\|$$
$$+\left[P_2\binom{0}{2}-2.P_1\binom{0}{1}\right] = +[1-2.0] = 1$$

But all the differences on the table 1 equal to one another; therefore, the constant K = 1.

C. That table only indicates the differences $+\left[P_n\binom{0}{n}-n.P_{n-1}\binom{0}{n-1}\right]$ having a plus sign; therefore, n is an even. Actually, n is an undetermined number. It can be either even or odd. So we must distinguish n into two cases.

$*$ n is even $\longrightarrow$ (n-1) is odd $\longrightarrow$ (n-2) is even.
$\downarrow$ $\downarrow$ $\downarrow$
$+[P_n(\overset{0}{n}) - n.P_{n-1}(\overset{0}{n-1})]$ $-[P_{n-1}(\overset{0}{n-1}) - (n-1).P_{n-2}(\overset{0}{n-2})]$ $+[P_{n-2}(\overset{0}{n-2}) - (n-2).P_{n-3}(\overset{0}{n-3})]$

$*$ n is odd $\longrightarrow$ (n-1) is even $\longrightarrow$ (n-2) is odd.
$\downarrow$ $\downarrow$ $\downarrow$
$-[P_n(\overset{0}{n}) - n.P_{n-1}(\overset{0}{n-1})]$ $+[P_{n-1}(\overset{0}{n-1}) - (n-1).P_{n-2}(\overset{0}{n-2})]$ $-[P_{n-2}(\overset{0}{n-2}) - (n-2).P_{n-3}(\overset{0}{n-3})]$

Thus, we get Table 2 containing the differences in whichever cases.

$$\pm[P_n(\overset{0}{n}) - n.P_{n-1}(\overset{0}{n-1})] = 1$$
$$\|$$
$$\mp[P_{n-1}(\overset{0}{n-1}) - (n-1).P_{n-2}(\overset{0}{n-2})] = 1$$
$$\|$$
$$\pm[P_{n-2}(\overset{0}{n-2}) - (n-2).P_{n-3}(\overset{0}{n-3})] = 1$$

- - - - - - - - - - - - - - -
- - - - - - - - - - - - - - -

$$-[P_5(\overset{0}{5}) - 5.P_4(\overset{0}{4})] = 1$$
$$\|$$
$$+[P_4(\overset{0}{4}) - 4.P_3(\overset{0}{3})] = 1$$
$$\|$$
$$-[P_3(\overset{0}{3}) - 3.P_2(\overset{0}{2})] = 1$$
$$\|$$
$$+[P_2(\overset{0}{2}) - 2.P_1(\overset{0}{1})] = 1$$

D. Based on Table 2, we deduce a list of $P_{n_i}\binom{o}{n_i}$ with n_i from 2 to n:

$$P_2\binom{o}{2} = 2.P_1\binom{o}{1} + 1$$
$$P_3\binom{o}{3} = 3.P_2\binom{o}{2} - 1$$
$$P_4\binom{o}{4} = 4.P_3\binom{o}{3} + 1$$
$$P_5\binom{o}{5} = 5.P_4\binom{o}{4} - 1$$

...

$$P_{n-2}\binom{o}{n-2} = (n-2).P_{n-3}\binom{o}{n-3} \pm 1$$
$$P_{n-1}\binom{o}{n-1} = (n-1).P_{n-2}\binom{o}{n-2} \mp 1$$
$$P_n\binom{o}{n} = n.P_{n-1}\binom{o}{n-1} \pm 1$$

In this list, we recognize $P_n\binom{o}{n}$, $P_{n-1}\binom{o}{n-1}$, $P_{n-2}\binom{o}{n-2}$ ... with the last term 1 having double sign (±), (∓), (±) ... because they depend on n, which is even or odd as we have explained at the establishment of Table 2.

With this list, we calculate $P_{n_i}\binom{o}{n_i}$ from $P_2\binom{o}{2}$ to $P_5\binom{o}{5}$ and write their value under form of factorial:

ANCIENT AND MODERN MATHEMATICS

$$P_2\binom{0}{2} = 2.0 + 1 = \frac{2!}{2!}$$

$$P_3\binom{0}{3} = 3.\frac{2!}{2!} - 1 = \frac{3!}{2!} - \frac{3!}{3!}$$

$$P_4\binom{0}{4} = 4.\left(\frac{3!}{2!} - \frac{3!}{3!}\right) + 1 = \frac{4!}{2!} - \frac{4!}{3!} + \frac{4!}{4!}$$

$$P_5\binom{0}{5} = 5.\left(\frac{4!}{2!} - \frac{4!}{3!} + \frac{4!}{4!}\right) - 1 = \frac{5!}{2!} - \frac{5!}{3!} + \frac{5!}{4!} - \frac{5!}{5!}$$

With these results, we recognize that all partial permutations $P_{n_i}\binom{0}{n_i}$ above compose of a series of fractions that possess three significant properties as following:

- The numerator of each fraction is a factorial of which the first factor matches the numbers of elements in $P_{n_i}\binom{0}{n_i}$, for example, $P_4\binom{0}{4} = \frac{4!}{2!} - \frac{4!}{3!} + \frac{4!}{4!}$ in which the numerator is 4! of which the first factor 4 matches the numbers 4 elements in $P_4\binom{0}{4}$.

- The denominators are the factorials that start from 2!, 3!, 4! ... and then end at a factorial of which the first factor matches the numbers of elements $\in P_{n_i}\binom{0}{n_i}$, for example, $P_5\binom{0}{5} = \frac{5!}{2!} - \frac{5!}{3!} + \frac{5!}{4!} - \frac{5!}{5!}$. We recognize that the denominators start from 2!, 3! ... and then end at 5! of which the first factor 5 matches numbers 5 elements $\in P_5\binom{0}{5}$.

- In the series of fractions of $P_{n_i}\binom{0}{n_i}$, we recognize that the sign of a fraction depends strictly on its denominator. The fraction of which the denominator is a factorial with the first factor is an even has a plus sign (+), and the fraction of which the denominator is a factorial with the first factor is an odd has a minus sign (−), for example, $P_5\binom{0}{5}$ has $\dfrac{5!}{2!}$ and $\dfrac{5!}{4!}$ because 2 and 4 are even and $-\dfrac{5!}{3!}$ and $-\dfrac{5!}{5!}$ since 3 and 5 are odd.

E. Because $P_2\binom{0}{2}$, $P_3\binom{0}{3}$, $P_4\binom{0}{4}$, and $P_5\binom{0}{5}$ possess the three properties above, so we can use the recurrent reasoning method to prove $P_n\binom{0}{n}$ also has these three properties. To do so, we suppose $P_{n-2}\binom{0}{n-2}$ still has the same properties of $P_2\binom{0}{2}$, $P_3\binom{0}{3}$, $P_4\binom{0}{4}$, and $P_5\binom{0}{5}$. That means:

$$P_{n-2}\binom{0}{n-2} = \frac{(n-2)!}{2!} - \frac{(n-2)!}{3!} + \frac{(n-2)!}{4!} - \frac{(n-2)!}{5!} + \ldots \mp \frac{(n-2)!}{(n-3)!} \pm \frac{(n-2)!}{(n-2)!}$$

F. Because $P_{n-1}\binom{0}{n-1} = (n-1).P_{n-2}\binom{0}{n-2} \mp 1$ [see the list of $P_{n_i}\binom{0}{n_i}$], we get:

$$P_{n-1}\binom{0}{n-1} = (n-1).\left[\frac{(n-2)!}{2!} - \frac{(n-2)!}{3!} + \frac{(n-2)!}{4!} - \frac{(n-2)!}{5!} + \ldots \pm \frac{(n-2)!}{(n-2)!}\right] \mp 1$$

ANCIENT AND MODERN MATHEMATICS

Or we can write:

$$P_{n-1}\binom{o}{n-1} = \frac{(n-1)!}{2!} - \frac{(n-1)!}{3!} + \frac{(n-1)!}{4!} - \frac{(n-1)!}{5!} + \ldots \pm \frac{(n-1)!}{(n-2)!} \mp \frac{(n-1)!}{(n-1)!}$$

Because $P_n\binom{o}{n} = n.P_{n-1}\binom{o}{n-1} \pm 1$, we can write:

$$P_n\binom{o}{n} = n.\left[\frac{(n-1)!}{2!} - \frac{(n-1)!}{3!} + \frac{(n-1)!}{4!} - \frac{(n-1)!}{5!} + \ldots \pm \frac{(n-1)!}{(n-2)!} \mp \frac{(n-1)!}{(n-1)!}\right] \pm 1$$

Hence, we deduce the general formula:

$$P_n\binom{o}{n} = \frac{n!}{2!} - \frac{n!}{3!} + \frac{n!}{4!} - \frac{n!}{5!} + \ldots \mp \frac{n!}{(n-1)!} \pm \frac{n!}{n!} \quad (n \geq 2)$$

6.2: Approximate Formula

In the second member of the general formula, n! is the common numerator of the series of fractions. So we can set n! as the common factor for the second member. Then we get:

$$P_n\binom{o}{n} = n!.\left[\frac{1}{2!} - \frac{1}{3!} + \frac{1}{4!} - \frac{1}{5!} + \ldots \mp \frac{1}{(n-1)!} \pm \frac{1}{n!}\right]$$

We recognize that the series of fractions in the brackets get the alternate sign from one to another and each fraction of the series gradually decreases from the first $\frac{1}{2!}$ to the last fraction $\frac{1}{n!}$, which vanishes in case n approaches to $+\infty$. So it is a convergent series when n tends to $+\infty$, the series tends to $\lim_{n \to \infty} \left(1 - \frac{1}{n}\right)^n = 1/e$. Therefore, when n is large, we get the approximate formula:

$$P_n \binom{0}{n} \approx n!/e \quad (e \approx 2.718281828) \quad (n \geq 2)$$

Formula Test

In the approximate formula, the larger the n, the more accuracy of e we get. But how large would n be in relation with e? To answer this question, we have to test the formula $P_n \binom{0}{n} = n!/e$ with n = 2, 3, 4, 5 ... to see how it works:

- **n = 2.** We get:

  $P_2 \binom{0}{2} \approx 2!/2.718 \approx 0.736$ (e≈2.718)

  Because 0 < 0.736 < 1 and $P_2 \binom{0}{2}$ is a positive integer, so $P_2 \binom{0}{2}$ must equal to the upper integer 1 (1 is closer to 0.736 than 0). $P_2 \binom{0}{2} = 1$

- **n = 3.** We get $P_3 \binom{0}{3} \approx 3!/2.718 \approx 2.208$. Because 2 < 2.208 < 3 and $P_3 \binom{0}{3}$ is a positive integer, so $P_3 \binom{0}{3}$ must equal to the lower integer 2. $P_3 \binom{0}{3} = 2$ (true).

- **n = 4.** We get $P_4 \binom{0}{4} \approx 4!/2.718 \approx 8.830$. Because 8 < 8.830 < 9, so $P_4 \binom{0}{4}$ must equal to the upper integer 9. $P_4 \binom{0}{4} = 9$ (true).

- **n = 5.** We get $P_5\binom{0}{5} \approx 5!/2.718 \approx 44.150$. Because $44 < 44.150 < 45$, so $P_5\binom{0}{5}$ must equal to the lower integer 44. $P_5\binom{0}{5} = 44$ (true).

G. We recognize that the accuracy of partial permutations $P_n\binom{0}{n}$ depends on the accuracy of e in the approximate formula. If n is larger, e must get more decimal places. All the previous tests with n = 2, 3, 4, and 5, we only used $e \approx 2.718$. The results of the tests may be incorrect if n becomes larger such as n = 8, 9 ... Let's take a test with n = 8 to see what happens:

$$P_8\binom{0}{8} \approx 8!/2.718 \approx 14{,}834.437$$

Because $14{,}834 < 14{,}834.437 < 14{,}835$, so $P_8\binom{0}{8}$ must equal to the lower integer: 14,834. $P_8\binom{0}{8} = 14{,}834$ (unusual). But the general formula gives us the exact result:

$$P_8\binom{0}{8} = \underbrace{\frac{8!}{2!}}_{20{,}160-} - \underbrace{\frac{8!}{3!}}_{6{,}720+} + \underbrace{\frac{8!}{4!}}_{1{,}680-} - \underbrace{\frac{8!}{5!}}_{336+} + \underbrace{\frac{8!}{6!}}_{56-} - \underbrace{\frac{8!}{7!}}_{8+} + \underbrace{\frac{8!}{8!}}_{1}$$

$P_8\binom{0}{8} = 14{,}833$ (true). Therefore, the test with e = 2.718 gives the false answer $P_8\binom{0}{8} = 14{,}834$. To get the true answer, we have to add more decimal places for e, such as e = 2.71828. Then we take a test again:

$$P_8 \begin{pmatrix} 0 \\ 8 \end{pmatrix} \approx 8!/2.71828 \approx 14{,}832.909$$

Because $14{,}832 < 14{,}832.909 < 14{,}833$, so $P_8 \begin{pmatrix} 0 \\ 8 \end{pmatrix}$ must equal to the upper integer $14{,}833$. $P_8 \begin{pmatrix} 0 \\ 8 \end{pmatrix} = 14{,}833$ (true). Now, we take a test with n = 9. $P_9 \begin{pmatrix} 0 \\ 9 \end{pmatrix} \approx 9!/2.718 \approx 133{,}509.934$

Because $133{,}509 < 133{,}509.9369 < 133{,}510$, so $P_9 \begin{pmatrix} 0 \\ 9 \end{pmatrix}$ must equal to the upper integer $133{,}510$: $P_9 \begin{pmatrix} 0 \\ 9 \end{pmatrix} = 133{,}510$ (unusual). But the general formula gives us the exact result:

$$P_9 \begin{pmatrix} 0 \\ 9 \end{pmatrix} = \underbrace{\frac{9!}{2!}}_{=181{,}440-} - \underbrace{\frac{9!}{3!}}_{60{,}480+} + \underbrace{\frac{9!}{4!}}_{45{,}120-} - \underbrace{\frac{9!}{5!}}_{3{,}024+} + \underbrace{\frac{9!}{6!}}_{504-} - \underbrace{\frac{9!}{7!}}_{72+} + \underbrace{\frac{9!}{8!}}_{9-} - \underbrace{\frac{9!}{9!}}_{1}$$

$P_9 \begin{pmatrix} 0 \\ 9 \end{pmatrix} = 133{,}496$ (true). Therefore, the test with e = 2.718 gives the false answer, $P_9 \begin{pmatrix} 0 \\ 9 \end{pmatrix} = 133{,}510$. To get the true answer, we take e = 2.71828. We test again. $P_9 \begin{pmatrix} 0 \\ 9 \end{pmatrix} \approx 9!/2.71828 \approx 133{,}496.181$. Because $133{,}496 < 133{,}496.181 < 133{,}497$, so $P_9 \begin{pmatrix} 0 \\ 9 \end{pmatrix}$ must equal to the lower integer:

$133{,}496$. $P_9 \begin{pmatrix} 0 \\ 9 \end{pmatrix} = 133{,}496$ (true)

Remark

Throughout the tests with the proper e, we recognize two important cases:

- n is an even. $P_n \binom{o}{n}$ always equals to the upper integer such as: 14,832 < 14,832.909 < 14,833. $P_8 \binom{o}{8}$ = upper integer 14,833. In the event of $P_n \binom{o}{n}$ equal to the lower integer, that is due to the improper e or the error of the calculation.
- n is an odd. $P_n \binom{o}{n}$ always equals to the lower integer such as 133,496 < 133,796.181 < 133,497. $P_9 \binom{o}{9}$ = lower integer 133,496. In the event of $P_n \binom{o}{n}$ equal to the upper integer, it's due to the improper e or the error of the calculation.

Comment on General and Approximate Formula

There are two formulas to determine $P_n \binom{o}{n}$ in different ways, so we must select the most suitable case for them to get their full advantages.

General Formula

This formula is independent of the base e. It gives us the exact numbers of permutations contained in $P_n \binom{o}{n}$. But using the general formula, it takes more work and time than using the approximate formula. However, in case n is a large number and we don't get e on hand, we have to use the general formula or do nothing.

Approximate Formula

In most cases, this formula solves $P_n \binom{o}{n}$ with ease and rapidity. But it depends strictly on the base e. In case n is a large number, to get the job done, we have to get e on hand. Otherwise, we must use the general formula or take a lot of time to have e.

Application

Problem I

There is a group of seven children. Each child has his own toy. Because they have enjoyed their own toys for a long time, they don't take any pleasure in their toys anymore. Therefore, they would like to exchange their toys in such a manner that no one keeps his own toy, but instead gets a new one. How many ways are there for these seven children to perform the satisfactory exchanges?

Solutions

If after the satisfactory exchanges, no child keeps his own toy anymore but gets a new toy, that means the toys are exchanged from one to another. Therefore, the children are equivalent to positions; the toys are equivalent to the transposed elements of the partial permutations $P_n \binom{0}{n}$. Here we get n = 7 children with seven toys and the total ways of exchanges represented as $P_7 \binom{0}{7}$. There are two solutions:

1. In case we don't have e on hand or don't renumber e, we apply the general formula of $P_n \binom{0}{n}$ to get the answer:

$$P_7 \binom{0}{7} = \underbrace{\frac{7!}{2!}}_{=2{,}520-} \underbrace{- \frac{7!}{3!}}_{840+} + \underbrace{\frac{7!}{4!}}_{210-} - \underbrace{\frac{7!}{5!}}_{42+} + \underbrace{\frac{7!}{6!}}_{7-} - \underbrace{\frac{7!}{7!}}_{1}$$

   So we have the answer:

   $P_7 \binom{0}{7} = 1{,}854$ ways

2. In case we have e = 2.718, we apply the approximate formula to get the answer: $P_7 \binom{0}{7} \approx 7!/2.718 \approx 1{,}854.305$. We recognize:

   $1{,}854 < 1{,}854.305 < 1{,}855$

   Here n = 7 is an odd, so $P_7 \binom{0}{7}$ is equal to the lower integer 1,854. The answer is:

$P_7\binom{0}{7} = 1{,}854$ ways.

Problem 2

Considering the partial permutations $P_8\binom{2}{6}$, find the numbers of single permutation $\in P_8\binom{2}{6}$.

Solutions

According to Theorem 1, we get:

$$P_8\binom{2}{6} = {}^8C_2 \cdot P_6\binom{0}{6} = \frac{8!}{(8-2)!2!} \cdot P_6\binom{0}{6} = 28 \cdot P_6\binom{0}{6}$$

Like problem 1, there are two solutions:

1. In case we don't remember e, we apply the general formula to find

$$P_6\binom{0}{6} = \frac{6!}{2!} - \frac{6!}{3!} + \frac{6!}{4!} - \frac{6!}{5!} + \frac{6!}{6!}$$

   $P_6\binom{0}{6}:$
   $= 360 - 120 + 30 - 6 + 1$
   $P_6\binom{0}{6} = 265$

   Therefore, we get the answer:

   $P_8\binom{2}{6} = 28 \times 265 = 7{,}420$ single permutations.

2. In case we remember e = 2.718, we apply the approximate formula to find $P_6\binom{0}{6}$: $P_6\binom{0}{6} \approx 6!/2.718 \approx 264.901$. Because 6 is an even, so $P_6\binom{0}{6}$ is equal to the upper integer 265. We deduce the answer:

$P_8\binom{2}{6} = 28 \times 265 = 7{,}420$ single permutations

www.ingramcontent.com/pod-product-compliance
Lightning Source LLC
Chambersburg PA
CBHW031838170526
45157CB00001B/345